The Ancient War Game

THE ANCIENT WAR GAME

CHARLES GRANT
(author of *The War Game*)

ST. MARTIN'S PRESS
NEW YORK

Dedication

To my wife, son and daughter, in appreciation of their understanding and encouragement.

 For information write:
St. Martin's Press Inc., 175 Fifth Avenue, New York, N.Y. 10010

Printed in Great Britain

Library of Congress Catalog Card Number: 74-79959
First published in United States of America in 1974

Contents

List of Illustrations

PHOTOGRAPHS

MAPS

Preface

To cover in any detail the entire military history of the ancient world and how one may go about wargaming in its context would be an impossible task outside a volume—or volumes—many times larger than this present one, and it is more than a little saddening to realise just how much has had to be omitted. I have attempted, in the light of my own experience of what is most popular in the field, to describe, if only superficially, what makes wargaming in the ancient period so prodigiously intriguing. It is correct to say that, since committing myself fairly substantially to the hobby in this historical framework, the pleasure I have had from it has been truly enormous. Being as gregarious as the majority of wargamers, meeting either in combat or in conversation so many individuals of similar tastes and like mind has been a constant pleasure as well as forcing me to keep up with current ideas and trends, so marked with such an array of keen young intellect participating. In many visits to centres of ancient wargaming it has been most rewarding to find such lively interest shown by such a cross-section of players of every occupation and profession and of every age from sixteen to sixty. To them, and especially to those members of my own immediate circle with whom I have battled for upwards of three years with practically every possible army and on every conceivable sort of battlefield, I am really deeply grateful for the pleasure I have derived from the hobby, and to any convert I need only wish but a fraction of the satisfaction I myself have experienced from it.

I

Introduction

Second only to the rapidity of growth of the general hobby of wargaming itself in recent years has been the truly remarkable increase in the numbers of those whose interest lies in the ancient world, in the far distant days of Greece and Persia, of Rome and Carthage, and of Egypt and Macedon—in this sphere the proliferation of wargamers who do battle with hoplites and legionaries or with chariots and war elephants is quite astonishing. Time was when, in terms of popularity, the Napoleonic 'épopée' was away ahead of any other period; indeed old hands, in whose ranks I have to enrol myself, will recall the days when it was the only epoch to have any appeal at all. Happily, enlightenment and progress are powerful influences and today it is pleasant to observe that, although comparison figures are naturally not available, the 'ancient' fans are not far short of the Napoleonics in mere numbers and bid fair to outstrip them in organisation and communication.

It might be a good thing to ask why this should be, and the answer is by no means easy to discover. Inter alia, it may well be that warfare, now seemingly a rather unlikely contingency on a major scale, can be considered fairly remote and so can be studied without any immediate personal connotation, and ancient historical studies can be re-enacted without risk of an interested party's being regarded, if not as a kind of warmonger, at least as a harmless eccentric. This theory might be coupled with the fact that there has been a general increase in leisure time, with the consequent search by increasing numbers of people for a spare-time occupation which will at once provide a modicum of interest and afford a degree of mental stimulus as well. It goes without saying, on the strictly practical side, that much is also due to the manufacturers who provide the miniature soldiery without which there can really be no war game and to the numerous retailers who stock such products for sale to a grateful wargaming public.

Of course, quite apart from all this, the basic question of why be a wargamer is a most difficult one to answer. It has been put to me many times in a variety of contexts and in many different forms, but I have never been able to provide a reasonable reply satisfactory either to the questioner or to myself. One can simply claim to have been 'bitten by the bug', like any other hobbyist, but this is not really complete in itself. There are all sorts of contributory factors—a liking for history is one that immediately springs to mind. There may also be the awareness that, like it or not, man is a combative animal. For many centuries, the ultimate issue between contending nations has been a recourse to the force of arms and, harmlessly upon a war games table, such contests can be studied and refought bloodlessly in miniature. Added to these is by no means the least important factor, the aesthetic appreciation of colour and panoply and the stirring spectacle of formed bodies of soldiers, whether this be in actual form or in miniature. Who could have watched without some form of appreciation the television showing of the tremendous military parade to celebrate the 2,500th anniversary of the foundation of the Persian Empire as it took place in the autumn of 1971? The marching bodies of infantry, cavalry and chariots, all in the clothing and accoutrements of so many thousands of years ago, was calculated to provide an enormous fillip to every ancient wargamer who saw them. Just as important too is the realisation that while taking part in a war game one is pitting one's wits against those of an opponent in an exercise whose rules are framed to conform with the movement and weaponry inherent in any historical military situation.

All these things, and indeed many others, must combine to provide the wargamer's raison d'être, and they are all splendidly fulfilled by what we can describe as the 'ancient period'. Even at this early stage it might be as well to define, if only in general terms, just what we mean when we talk largely about the ancient period. This is not a difficult problem and we can say that ancient wargaming is bounded on the one hand by 1500 BC and on the other by 500 AD. Before the former year it can hardly be claimed that organised warfare between states took place to any degree at all. It can generally be assumed that weaponry had scarcely progressed beyond the stone club stage and, even in the first few centuries of the time we have laid down, there was but a very limited number of peoples able to field anything like organised and trained forces. Later than 500 AD, however, we are approaching the beginnings of the Middle Ages, and 500 AD is a suitable time, if not in all aspects strictly accurate, to divide the 'ancient' from the 'medieval'.

One thing will be readily apparent to the keen eye of the student. This is of course the simple fact that our period consists of no less than 2,000 years, in effect twenty centuries of technical, social and military development, and at once there becomes plain one particular attraction which might be calculated to influence the potential wargamer to adopt the period—the truly enormous variety of armies and military systems from which to make a choice. Consider some of the generally accepted and most popular periods: the Napoleonic, for instance, lasted a mere twenty years or so; the American Civil War about half a dozen years; World War II the same; and indeed if we are generous and take the entire horse and musket period (the epoch of the cavalry charge and the smooth-bore musket and cannon), we begin at around 1700 and halt at, say, 1850—a mere hundred and fifty years, a drop in the ocean of time as compared with our 2,000. And what a wealth of fascinating material awaits our attention there! It used to be said by those who knew no better that ancient wargaming must be less interesting than that based upon more recent times, with the armoured fighting vehicles, the long-range guns, the aircraft and the plethora of technical advances which have so fundamentally altered the course of modern warfare. Nothing could be further from the truth than the suggestion that ancient warfare is now too simple to be considered worthwhile. After all, from 500 AD to the present is five centuries less than the 2,000 years encompassed in our ancient period, and what fundamental and startling changes this lengthy span produced on the military scene in the improvement and adoption of weapons and in the techniques of warfare. Even the names of the generals speak for themselves—Hannibal, Alexander and Caesar are but a single triumvirate of leaders whose equal it would be hard to find in more recent days.

From the dawn of history when the troops of Egypt marched with glittering chariots and bronze weapons to do battle with the Hittites, we can make an enthralling pilgrimage in time, studying the highly developed military art of that warrior people, the Assyrians, examining the campaigns of the Great Kings of Persia and their defeats at

Slingers were extensively used in many parts of the ancient world; here they are in action on the slopes of a convenient hill

the hands of their Greek enemies, watching the Roman legionaries as they march north into Gaul and east into Asia, and following the Gothic cavalry in their plunge into Europe. There is all this and more—the subtleties and skills of siege warfare, encouraged by such dramatic characters as Dionysius of Syracuse, and the practical abilities of professionals like Miltiades and Xenophon are only two of the facets we can study and strive to emulate on the war games table. Truly, the ancient period has so much to offer in the way of interest, as well as—dare we say it?—romance, that it seems to be the perfect epoch for those who wish to delve into the strategy and tactics of the Great Captains of the distant past and to reproduce in miniature the achievements of their favourite leaders and chosen armies, whether it be the horsemen and archers of the Persians, the dogged legionaries of Rome, the elephants of Hannibal or the chariots of Ashurbanipal of Assyria. They are, each and every one, ready to be marshalled on a war games table, to be manoeuvred, advanced or retired as fancy or the situation merit, or let us piously hope, the tactical sense of the player dictates.

Of course, as well as all the historical advantages involved in wargaming in an ancient context, there have been other reasons, as has already been indicated, for the growth in this phase of the wargaming hobby as a whole, not the least of these being the availability of miniature figures specifically designed for use in war games. It is assumed, incidentally, that the enthusiast will wish to employ figures rather than that he should content himself with coloured blocks, counters or such similar monstrosities. In recent years, the market has been deluged with war game figures of all sorts of periods and of diverse quality, but for some time the ancient player suffered in that he was offered very little of any use to him except the 'flat' figures emanating from the continent, usually in fact from Germany, home of the *zinnfigur*, the tin soldier. These 'flats' (actually two-dimensional, highly engraved types) can be of a very high quality and those produced by early moulds, now lost or destroyed, are extremely valuable and much sought after by collectors. When painted with skill and patience and viewed en masse, particularly in dioramic settings, they provide a vivid and impressive spectacle, although when seen 'end on' they naturally suffer somewhat. Despite the undoubted high standard of production, they have never achieved very much popularity in this country compared with the three-dimensional models, graphically described by the French as 'ronde bosse', which were originally produced in England by Britains Ltd. Alas, they are made no longer, at least not in metal. Many wargamers of my generation and later will look back on early war games, possibly a shade rudimentary by present-day standards, where rows of Britains' soldiers were mown down by matches fired from miniature cannon! Enough of nostalgia, though, for it is the present with which we are concerned, and now the ancient wargamer is as fortunate even as his Napoleonic colleague, and recruits are available for mustering into the ranks of pretty well every ancient army one can imagine, and in what is now generally

Cavalry of the ancient world; a body of cataphracts, with horse and man both armoured

accepted (as far as anything is accepted by the intensely individualistic wargamer) as a standard size. This is in fact the 25mm scale, which briefly means that the foot soldier, the infantryman, measures just that from base to apex, with the cavalryman proportionately larger. It is true that different manufacturers have rather varying ideas on just what constitutes 25mm, a seemingly rather elastic term, and the wargamer will find figures ranging between 20mm and 28mm, depending upon which firm's products he purchases. However, more will be devoted to the subject of manufacturers at the proper time and it is upon this 25mm figure that our wargaming and its rules are based, this being a most important and indeed a vital element.

It is always recognised that one of the most interesting features of wargaming, and indeed of any study or participation in things military, is the research that can be carried out in connection therewith. Much literature, both learned and popular, is available for the student who wants to study either the battle tactics of the time or the more intimate details of equipment and uniform of former days. In connection with this, it is interesting to note that alone of the various periods of military history which can be used for the staging of war games, the ancient one is alone in having its own organisation, society and publication, and this will also be considered in the appropriate place. Complete newcomers, however, might be advised to study appendices III and IV before going in to the main text of this book.

Having furnished, it is hoped, a very brief background to our subject, it remains but to discuss its various aspects in more detail and primarily to say something about the rules which govern the playing of the ancient war game; how one proceeds to set up a battlefield and just how the mechanics of the game provide for the solution of practical tactical problems as well as speculative situations. It will be my pleasant task to throw some light upon these things, illustrating the principles laid down with examples from actual war games in which I have participated or which I have observed over a not inconsiderable number of years.

II

First Principles of the Rules

At this stage there is no need to make a really detailed examination of war game rules other than to note that, as in the case of most war game periods, varying conceptions of seemingly basic principles are to be found, and that wargamers have frequently quite startlingly different interpretations of fundamentals over which there would appear to be no room for argument. It could indeed be said that one great advantage of wargaming as a pastime is that one can draw up one's own code of rules and put it forward as being the best in its field. What can happen is that some enthusiast will uncover some source material or reference work which, he claims, proves conclusively that, for instance, a certain cannon had a maximum range of x yards and not the y yards as had previously been ruled. This sort of thing tends to cause confusion in the ranks, particularly if *A* and *B* meet for a wargame and, to their mutual consternation, find that one deems x to be correct while the other adheres firmly to the y doctrine. It must be pointed out that, in sharp contrast to this state of affairs, the ancient wargaming world is especially fortunate in that the particular rules used for games set in the period are used almost universally, and a wargamer can therefore travel from one area to another in the knowledge that he will not have to assimilate an entirely new and different set of rules—with which he might well disagree—before engaging in an encounter with a 'local'. Rather than overtaxing the mental digestive processes of the reader with the administration of large and unpalatable doses of tabulated movement rates, weapon capabilities and the like, this chapter will expand a little on the main principles upon which they are considered and how they operate.

These principles are not too intimidating, are relatively few in number, but are quite vital for the achievement of a successful and realistic war game. In brief, we must discuss in the first place ground scale and its corollaries (deployment and movement rates), the effect of different weapons and the troops using them upon varying sorts of enemy, and last, and as important as any, morale.

It is very apparent even to the neophyte that, faced with a war games table large or small he will have to determine just what this area purports to relate to in terms of real life distance. He will naturally see that, if his table is square, with a side of 5 feet, he can, by changing the ratio of table size to real distance, make the five feet easily represent 50 yards, 500 yards, or indeed 5 miles. All these scales can be relevant, depending upon whether the aim is to reproduce a small action, extending over a couple of thousand square yards or so of terrain, or a major battle covering several square miles with the equivalent of several thousands of men taking part. For our purpose we shall normally adopt a single ground scale, one whose suitability has been well proven by a great number of war games, and this is a very simple one, to wit 1″ equalling 10 paces. In the far-off days of our period yards, or indeed metres for that matter, did not exist, so we shall ignore these modern measurements and retain the one which has been constant as long as men have marched upon two legs, this being so whether the distance was determined in Gaul, Greece or Assyria. If then we take the pace to be around 30″, give or take a little, we can obtain reasonably accurate historical distances which will fulfil all our requirements. Should one consider that the 'pace' idea obscures the issue or is a trifle pedantic, one can forget it and substitute yards or metres. The substitution will have no effect on relative movement rates or weapon ranges.

This ratio of 1″ = 10 paces is a cornerstone of our war game edifice, second only to it being the time element, or what we describe as a war game 'move' or 'period'. The latter term is preferable as, during the time interval chosen, there need not in fact be any movement at all, for troops do stand fast from time to time and, if we read our military memoirs correctly, they did so for long and tedious periods. In effect, our war game progresses in a series of periods (one occasionally hears of these being described as 'bounds') and again, without going into too much detail, it has been found that one such period will most conveniently be reckoned as half a minute. When we speak of a period, therefore, we are referring to an interval of time, and all the movement, discharge of missiles, hand-to-hand fighting or other military activity which took place therein. The war game, by virtue of this, consists of a number of periods during each of which troops are moved the appropriate distances if so desired and the result of any sort of combat is assessed. Periods for the opposing players are naturally simultaneous, since for obvious reasons this is a much more realistic system than any which involves alternate moves. It means in effect that continuing action is 'frozen' every 30 seconds—the process is similar to what happens when a film is stopped at regular intervals, leaving the actors in a sort of suspended animation.

Now, with both ground and time scales adopted we have made considerable progress and our next task is to determine just what moves can be made by our war game figures in the period we have chosen. Naturally certain types of soldier—the lighter and more active—will get about much more quickly than those who are heavily armoured, while mounted men will obviously be even more speedy. In laying down what is deemed to be the correct moves it must not be forgotten that speeds in practice are not always as great as might be thought. We are after all dealing with the movement of substantial bodies of men who must manoeuvre in well-ordered lines and who must preserve their 'dressing' in the manner of disciplined soldiers in any period of history. One example should suffice to illustrate the point.

An average man, marching with some slight degree of urgency, as in the process of executing some military movement, and not too heavily laden with arms and equipment, will progress at a rate of some 120 paces to the minute, and so in half that time will cover a distance of 60 paces. Ergo, scaling this distance down as we have already described, a miniature soldier on a war game table will move the equivalent of 60 paces or, in our scale, 6″. Thus, in simple terms, we enact that the basic infantry move is 6″. This will obviously be subject to some alteration in the case of men more heavily burdened, 'heavy' infantry, in other words. Two examples of such are the Greek hoplite with cuirass of metal, helmet and heavy shield, and the later Roman legionary with helmet, banded armour (the 'lorica segmentata'), shield and heavy throwing spear (the 'pilum'). The movement rate for such troops will consequently be less, something in the order of 40 paces per half minute, giving us a heavy infantry move, in effect, of 4″. In contrast, the lightly or totally unarmoured skirmisher —the traditional light infantryman—nipping about throwing javelins or using bow and arrows will be less impeded and will be able to get about much more quickly. The light infantry move will therefore be one of 8″. All these types, of course, will when pressed or when in the act of carrying out a charge be able to run for a short distance at the maximum speed consistent with maintaining the necessary formation. Thus the medium infantryman will have a 'charge move' of 8″ as opposed to his normal one of 6″. Other types will have an upgraded charge move in the same way. Cavalry moves will obviously be greater than those of infantry, but will be similar in that light cavalry will move more rapidly (i.e. they will have a greater move on the war game table—in actual fact one of 16″) than heavy cavalry. The latter make an ordinary move of 8″, having heavier horses bearing armoured riders.

There is yet another important consideration relating to distance, this being the range of missiles, arrows, slingstones and javelins. All these can be scaled down in just the same way as before. Taking the bow as our first weapon, we know of course that there were different types with varying capabilities. The simple self bow, made from a single stave of wood by barbarian tribesmen and so on, had an effective range much less than the powerful eastern composite bow built up of layers of horn, glue and wood. However, rather than overcomplicate the issue (bow ranges can easily be altered by the wargamer desirous of such refinements) we shall in the present instance opt for one general bow range for all types. This would in actual fact average something like 200 yards—this is the effective range—approximating to 240 paces, or 24″ on the wargames table. As we shall see it is less lethal at long range than it is when the target is in greater proximity to the archer. The hand-hurled javelin naturally has a much shorter range than the bow, and we make this 4″, while the stone or 'bullet' (the specially cast leaden ball) from the sling will be effective up to 12″. All this, it goes without saying, is the result of much research and very considerable war game experience which suggests that they are accurate as can be, and contributes largely to the war game's having the correct 'feel', one of the most important desiderata concerning the hobby. The javelin mentioned is, by the way, the light variety, several of the heavier throwing spears having such a limited range that its effect has to be incorporated in that of hand-to-hand fighting, or mêlée. Such is the case with the Roman 'pilum' and indeed with the Germanic 'francisca' or throwing axe, which seem to have had an extreme range of some 30 paces and which were hurled or thrown just before the legionary or the tribesman moved into hand-to-hand combat with close-fighting weapons.

If we proceed with the discussion of the principles we have specified we find that dispersion and deployment of troops, and consequently of war game figures, are closely connected with the operation of the ground scale. In short, it is generally accepted, bearing in mind that we shall be using 25mm figures, that it will be a most suitable ratio if we take one war game figure to represent 20 actual men, this being very satisfactory also from the visual point of view, a highly desirable feature. At this stage it is essential to point out the one figure/twenty men concept is a purely arbitrary one and is not the basis of the rules, this being, to repeat the dictum, that 1″ equals 10 paces.

Light/medium, heavy and light infantry, showing the dispersion appropriate to each type

To proceed, our regiment, cohort, phalanx or whatever will consist of a number of men covering a particular area, and the density of men in the formation in question will have an important bearing on the area they occupy. Thus, if we adopt the suggestion that one figure represents twenty men, this figure should cover, on the war game table, an area equating to that occupied in real life by twenty men. If these are in close order, virtually shoulder to shoulder as was the case sometimes with a Macedonian phalanx, then the area covered by the figure representing twenty men would be much less than if they were in a more open order, or if they were light or skirmishing troops whose function dictates that they be very much further apart than close-order infantry. To give a simple comparison—three medium infantry figures would in close order have the same frontage as only one light infantry figure. In other words the close-order infantry are in a formation three times denser than the one in which their lighter brethren operate. The same applies to mounted troops—light cavalry will be much more widely spaced than their 'heavy' comrades who must ride as closely together as possible. An accepted method to cover all this is for the war games figure to be fastened or stuck to a thin card base of predetermined size. Thus if we use the system as suggested by the Rules of the War Games Research Group (q.v.), a medium or heavy infantry figure will be on a base with a frontage of between 10mm and 15mm, while one light infantry figure will have a base with a 30mm frontage (the latter would be archers, slingers and the like). By doing a little calculation, it will be seen that a front of 40 paces—4″ on the table—would be occupied by something like 10 medium or heavy infantry figures, while the same number of light infantrymen would have a front of about 12″—this giving a graphic picture of the contrast between the deployment of close order infantry and that of light. As we shall see, the dispersion of the light infantry has a marked effect on their vulnerability to missile fire but more of that anon.

Illustrating troop formation density; open order light and close order heavy cavalry

It might be as well at this point to digress slightly on how we categorize for war game purposes the troops we have so glibly referred to as heavy, light and so forth. Much will depend upon the weight of weaponry and equipment and the tactical purpose of the type of troops, but generally we can take it that heavy troops, both horse and foot, wear metal armour and usually bear shield and helmet. All sorts of types come to mind, the Greek hoplite, the late Roman legionary and the Assyrian with ankle-length metal reinforced robe. Medium infantry would most probably have leather protection, or possibly none, but would carry an especially large shield, while light infantry, normally missile men, would be totally or almost entirely without armour, although slingers are usually shown as having a small shield which they would use even when in action. These categories are far from being rigid, since they could and frequently did overlap, and from time to time we find some unit or formation (Alexander the Great's 'hypaspists' are one such) for whom paucity of contemporary writing fails to provide an adequate description. In the same way, cavalry can be described as heavy and light, ranging from Alexander's Companions to the light mounted Scythians of the steppes. In addition to the former there are what can only be described as 'super-heavy' horsemen, with mount and rider both armoured, examples being Parthian and Sassanid Persian cataphracts and 'clibanarii'. More detail concerning all of them will be given at the proper place.

III
Combat and Casualties

In most war games (for instance those set in the Napoleonic Wars, American Civil War and Seven Years War) when two armies meet in combat, the players are aware with some exactitude of just what sort of army will be met, what sort of tactics will be employed and what kind of weapons will be used. Each will know that the enemy 'army' will be almost the precise counterpart of his own. If a wargamer fields a French army of 1800 to meet an Austrian army of 1815 there will be an absolutely minimal difference between the weaponry and the military capabilities of the two forces involved. The same will apply should Federal troops of 1861 meet a Confederate army of 1865 and if a Frederician army of 1742 clashes with an Austrian one of 1760. The point is that war game rules for the periods referred to are simple to construct in so far as they apply equally to both sides and there is already a built-in similarity. This can be the case in the ancient period when, say, a Roman army meets a Contemporary Carthaginian one, or when Greeks meet Persians—each battle would be fought in a time-setting common to both; but it might be as well to remember the extent of our period at this stage. Suppose one player boasts an early Egyptian army and his antagonist fields, just to give as extreme a contrast as possible, one consisting of Germanic tribesmen organised in warbands! Such a battle, by reason of geography as well as of time, could never have taken place and is indeed completely hypothetical. It is however one of the virtues of ancient wargaming that such speculative encounters on the war game table are made completely possible, and produce enormous interest and indeed excitement.

How then can such an exercise be organised? A comment on the question is simply that it is frequently done with enormous success and the system which permits this is far from being an obscure or esoteric one. To begin with, if we study our period in its entirety it can be seen that certain military systems are much superior to many others. At once the Romans come to mind, but the legions could be defeated and indeed were on more than one occasion—by the Germans, by the Carthaginians and by the Parthians—so obviously there were occasions when other weapons and tactics proved their superiority. Throughout the Ancient World and at all stages of its existence, there were great differences in military training and discipline, and the two thousand years we have spoken of did produce tremendous improvements in military science and technology. What is to happen then when, as we have suggested, a player fields an ancient Egyptian army against an opponent deploying Germanic warbands? Similarly, how can we cope with a battle between Assyrians and Ancient Britons? The problem might seem to be a considerable one, but it is really not so presupposing some research into the subject and a little constructive thinking. Although engagements such as we have mentioned could never really have occurred, tremendous delight is to be found in restaging them as though we were masters of time and space. How splendid it is to watch the discomfiture of a 'Roman' enthusiast, who thinks of his legions as supreme masters of war, when he sees them tumble into ruins before the onslaught of a hoplite phalanx from the city states of Greece! The problem involved is simply the technical one of achieving some sort of balance between forces which never met and whose military prowess in respect to each other is consequently unknown, with the important proviso that the aim should be to ensure some kind of equality between such speculative antagonists so that no player shall, as a matter of course, emerge as an inevitable victor. The problem is solved thus.

First, one needs some sort of common denominator for the fighting value of all sorts of troops, irrespective of period or nationality, having regard to their valour, weaponry, discipline and training. This can be done without a great deal of trouble. Take as a couple of examples, from opposite ends of the military scale, a semi-naked Pictish savage without armour of any sort but imbued with a natural courage and ferocity—a near-fanatic, possibly—and face him with a well protected, highly trained and disciplined Roman legionary, or maybe an 'Immortal' from the Imperial Household of one of the Great Kings of Achaemenid Persia, Darius or Xerxes. Each of these, considered in a context of skill in combat (natural or acquired), weaponry and innate courage, can be awarded a certain fighting value vis-à-vis each of the others, this being qualified by his mobility in action and by what he can do to defend himself and fellow-warriors, together with the circumstances in which he fights. It might well be that the effectiveness of the Pict against the Persian, with the latter's less potent weapons of short spear and wicker shield, would be greater than it would be against the Roman armed with heavy shield, helmet and armour and wielding heavy javelin and sword. One can arrive after some consideration at an individual value and thereby at a value for a unit—a simple matter of multiplying the fighting factor arrived at by number of men in the unit. Thus, when Picts meet Romans—their numbers being similar—the Picts might inflict x casualties upon them, while the Romans would probably be more successful, achieving $2x$ casualties; the Picts might cause $2x$ casualties among the Immortals, receiving the same themselves. Thus we get a standard of effect which *A* will normally have on *B*, *B* on *C*, and *C* on *A*, etc, this being derived from an extensive study of military history, the assessment of battle casualties and as much research as possible into all relevant factors.

Naturally numbers play an important role in what determines the outcome of any sort of combat. Although Romans will properly have a straightforward fighting advantage over Picts or similar warriors, it is apparent that 10 Romans will not usually have much joy against 100 Picts, but—and it is an important qualification—it could just happen that through vagaries of fortune, aided and abetted by an infinite number of surrounding circumstances, the 10 Romans could indeed on occasion send 100 Picts about their business. Here is the point where, to provide the swing on each side of the norm, a variable has to be introduced to allow for the inevitable element of chance in battle. This is the imponderable of war—a phrase I have used before and will indubitably use again. As in all well regulated war games dice are used to cater for this uncertainty factor. Two dice are thrown by each player, one die, say a black one, being deemed to be the 'positive', the other, possibly red, being the 'negative', and the result of this dice throw (in fact the difference between the numbers shown on the dice) is added to or subtracted from the standard fighting factor of the troops under consideration (if the number shown by one die is the same as that shown by the other, then neither addition nor subtraction will have to be made). The final weapon factor (including possible variable) and the number of men in combat will give the casualties inflicted upon the enemy. Now even with a very favourable 'chance' factor it is almost impossible for certain types of troops, usually barbarian in origin or wearing little or no armour, to defeat others of a heavier or more disciplined calibre. It can be done however and possibly a personal example might illustrate the point—which is that in warfare, as in most aspects of human endeavour, the unlikely can and did happen (that the combat involved had a favourable result for me is purely coincidental).

The incident involved a war game between myself, leading an army of Persians, and one Charles Tarbox, an enthusiastic Californian who during a visit to England in the not too distant past cut something of a swathe through home-grown wargamers and who in this instance commanded an army of Greeks—the period being about the fourth century BC. To defend a small hill, a fairly important part of my position, I had a body of javelinmen, unarmoured but carrying a small shield, against whom there came marching a strong body of Spartan hoplites with body armour, helmet and shield and wielding the characteristic long thrusting spear—very formidable adversaries who would normally have been expected to sweep away with some contempt my light infantry, despite the latter's having the advantage of fighting from higher ground. Naturally the standard fighting assessment of the Spartan hoplites against my javelinmen was considerably higher than was the reciprocal and the chances of the javelinmen were poor, particularly as they were also outnumbered. Indeed it must be said that my situation in this corner was a parlous one. However, when the two units met in a mutual charge and I rolled the two dice for the variable, I was fortunate enough to get the maximum positive which when added to the standard weapon value of the javelinmen and multiplied by the number engaged, produced considerably more casualties upon the Greeks than the latter inflicted, their 'commander's' dice-throwing being only moderate. The Spartans were not broken but had taken sufficient 'stick' to be pushed back on the next period—the distance involved was 3″ and equated with the unit's giving ground slowly under strong pressure.

In the next period (or move if you like) I again came up with an excellent positive variable once more driving the Greeks back 3″, and this was followed by yet a third successive 'push back'. When a unit has been pushed back three times in succession the probability is that it would continue this retrograde movement, and when this happens it incontinently breaks and runs. This the Spartans did, to my unconcealed joy and astonishment (it was indeed a most unlikely event) and to the patent disbelief of my American opponent, who however took the catastrophe with considerable sangfroid. I would be the first to acknowlege that this was a most unusual case, for in the vast majority of such conflicts the hoplites would have walked through such opposition with the minimum of bother. But this sort of thing could and *did* happen, and I consoled my 'enemy' by indicating a historical precedent of 300 BC when an Athenian general named Iphicrates with light troops destroyed a strong body of Spartan hoplites, to general amazement.

This then is briefly how a decision is arrived at when a hand-to-hand encounter takes place. Each unit, depending on its type, has a 'strike value' or 'weapon factor' varying with the opposition encountered which, used in combination with the physical circumstances—whether the unit is fighting downhill, is charging, or is in a disorganised state etc—and with the variable factor provided by the dice throw already described will give the total casualties inflicted on the enemy troops being fought. The number of casualties in actual war game figures can be determined by reference to rules. This system is especially valuable in our case, that of a very lengthy period wherein we stage battles which could not have occurred historically and where the decision has to be arrived at, in a manner of speaking, theoretically. In effect the determination of a commonly-based fighting factor for all sorts of troops throughout our time spectrum makes it very easy to match up the most unlikely opponents.

One very important consideration which should be spoken of is the fairly vital matter of the difference between the value, both as individuals and as a unit in combat, of regular troops (by which we mean paid men, trained and equipped by the state), and that of a non-regular or, for want of a better word, 'barbarian' force, by which we indicate fighting men who have no orthodox training and bear little or no allegiance to a central authority, being called to arms by their own chiefs. They receive no pay, ordinarily living by the acquisition of booty alone. The difference between the two is fairly self-evident, and the Roman legionary and the Germanic or Gaulish warrior can be taken as fairly typical examples of the two types. It costs nothing for the German or Gaulish chief to bring his warriors into the field, their fighting quality is innate if somewhat erratic, although their discipline in a strictly military sense is practically non-existent. On the other hand the Roman is paid, has had prolonged training and his arms and armour are provided by the State. This leads to the important point that when two armies are being assembled for wargaming purposes it would be unfair, if the ensuing battle is to start as an evenly matched one, for this to be done on a strictly numerical basis only. In effect, as the Romans are more costly to muster, they should be fewer in number than the Gauls, althought normally their fighting quality would be higher so thus we achieve a balance. This is the principle on which we operate. It is not strictly necessary for two contending war game armies to be exactly equal in fighting power (they rarely were historically) but for the purposes of the war game it is as well to start out on an equal footing, certainly if any sort of competition is involved. This is done by allocating a points value to each type of soldier, based upon his being regular or barbarian and upon an assessment of his fighting value. Whether troops are regular or barbarian, incidentally, has a significant effect on their morale, and this will be considered later.

Just as the fighting man in combat has a value appropriate to his type, so an archer, a slinger or a javelinman has a similar assessment based upon the effectiveness of the missile he shoots or hurls. Here again we have certain qualifications, not too many, but enough to simulate the real thing. Each missile weapon has a factor, a basic one for use when the maximum range of the weapon is involved, and one for shooting at over half the maximum range when a deduction is made. Let us say that the general weapon factor for archery fire is x. If the target is more than half the stated range for the weapon it is reduced by 1 and becomes $x - 1$. This relates to a stationary target, but let us suppose it to be on the move. It is more difficult to hit a moving man or men, so if during the period in which firing takes place the target has moved a certain specified distance—in effect it is travelling at a minimum speed—another deduction is made, again of 1. So should the target be at maximum range *and* moving, the factor will be $x - 2$. The basic factor itself is influenced by obvious things —it will be more difficult to incapacitate a heavily armoured man than one without protection, and by reason of their dispersion it will be more difficult to hit troops in open order than it would be were they shoulder to shoulder.

As with close combat, before we assess the number of casualties we have to add or subtract the variable to account for the vicissitudes of action—this again being the difference between two dice, plus or minus in effect. The theory behind all this may be explained by example. When a body of archers shoots at a target two extremes of result are possible—every arrow can hit a man or every arrow can miss completely. Generally however there will be an average number of hits and it is this which is catered for by the weapon factor or strike value, while the plus or minus dice throws give that slight variation above or below the norm which is due to all the multifarious causes we can imagine—the state of mind of the men in action, the effect of weather on their bowstrings, whether the men are hungry, tired or simply fed-up. This again reflects quite clearly the military imponderable and how it transforms the war game from a strictly mathematical exercise into something during which one must allow for the uncertain and largely unpredictable nature of warfare.

IV
Morale

Morale from the military point of view may be defined as being the state of mind of a body of soldiers with particular reference to their cheerfulness, their willingness to undertake some task, their readiness to endure losses in so doing and the extent of their confidence in their leaders. While good morale can occasionally be assumed to exist, it can never be predicted with complete certainty and no general can ever be one hundred per cent sure, if he orders some of his men to carry out some possibly perilous mission, that this will be done. In armies consisting in the main of highly disciplined, regular troops it might reasonably be expected that long training and confidence in an efficient officer corps normally ensures that orders will be obeyed and that the men carrying them out will give of their best while doing so. All this applies to our ancient period, but in the same era we find any number of examples of an army only a portion of which is made up of regular troops and quite a number of armies which are totally irregular or 'barbarian'. This does not mean, of course, that they are necessarily savages, but rather that they have never been subjected to any lengthy process of military discipline or indoctrination. Now it is most apparent that unwilling levies or pressed men are not customarily imbued with the desired martial spirit and that in certain circumstances they will be far from ready to carry out an order which might put them in jeopardy. Their reaction therefore to unexpected or adverse conditions might well be most unfavourable and not at all conducive to their carrying out any commands they might have been given.

To illustrate this, in a certain battle one of the opposing generals, having decided to carry out a wide flanking movement round his enemy's left wing, directed a body of irregular—or barbarian, if you like—light horse to move rapidly round the further side of a hill which dominated this section of the battlefield, and if circumstances proved favourable to do the enemy as much mischief as possible. Off went the cavalry, a unit of Scythian horse bowmen who were useful skirmishers but armed only with a bow and a smallish sword—not a great deal of use in a stand-up fight. Good progress was made until, swinging round the hill in question, the horsemen were confronted by a solid and most considerable array of infantry—two cohorts of Roman legionaries, armoured, disciplined and of a very high fighting quality, altogether too heavy metal for the Scythians who, seized with a sudden panic at the unexpected apparition, took to flight and made off with all speed. The Romans, in no way put out of countenance by the appearance of the cavalry, calmly stood fast and obeyed their orders which were simply to maintain this part of their general's line. The Scythians, had they obeyed their orders and skirmished with the Romans, could have safely inflicted considerable damage on them with archery fire as the Romans were without missile power at that moment. All in all, the example is a fairly classic one—if a trifle extreme—of how morale functions in different ways and how a momentary panic can remove all sense and reasoning ability from the minds of those affected by it.

Let me now reveal that this event did not occur in an actual historical battle but in a war game of which I was an interested spectator, and it illustrates how very accurately good war game rules cater for the intangible as well as the strictly material side of military operations. Let us examine briefly the case of the Scythians. They were probably a fairly volatile lot with no deep-seated loyalty to their army commander and they suddenly found themselves faced with Roman legionaries in considerable strength. They had no support, their general was not immediately present to enforce orders or provide inspiration, their armament did not appear sufficient (certainly not in close combat with such adversaries), their flanks were uncovered and all things considered they were not too happy in their total isolation. Consequently, when the customary assessment of the situation as it affected their morale was made, their reaction was to get speedily out of the way. Oddly enough, in contrast to what actually took place, these chaps could just have been seized with a lust to do battle and have gone tearing forward right into the Roman ranks. They did not, as we know, but the two extremes demonstrate just what we can find at opposing ends of the behaviour spectrum. Another example of this sort of thing is the case of a Germanic or Gaulish warband—as much an epitome of the barbarian ethos as one can imagine—whose chief or war leader might be struck down in action. The reaction of the band to this might be either to run like rabbits or to be seized with berserk fury and run screaming forward in a wild attempt to avenge their fallen lord. Both eventualities could and did happen historically.

In contrast to this unpredictable conduct is the behaviour of the trained soldier who, rarely if ever emulating the barbarian in rushing forward without orders to avenge the death of a leader or running away as a result of his demise, would stand fast and wait until the chain of command had been re-established. Only on the rarest of occasions will he be guilty of the extremes of action to which the barbarian is far more readily prone.

All this is leading up to the fact that behaviour patterns can be reasonably well foreseen with regular soldiers but not so much in the case of the non-disciplined warrior. Allowance must therefore be made in the ancient war game for such conduct as has been described as well as for the more disciplined reaction of the trained soldier. On many occasions, of course, the norm of expected conduct will inevitably be followed, the physical circumstances detrimental to the morale of some unit being counterbalanced by those beneficial to it. It will be the difference between these 'for' and 'against' circumstances which will play a large part in determining what will happen. Some of these circumstances will be very obvious. To begin with, the more support at hand or a number of friendly units close by will increase morale, as indeed will the sight of any friendly body of troops, no matter how far off it might be. If the unit itself or its friends are advancing, that will help—the general forward movement is stimulating to all. Is there support immediately to the rear or is there nothing but empty space behind? How about the flanks—are they secure? Then there are the more material points relative to our period—are the troops armoured or have they been provided with shields, for instance? Is there any friendly cover (woods, for example) near at hand? All these points and more can be added to give the positive or plus side of the actual physical factors governing morale, and the same things can be counted on the debit or minus side. How many enemy troops are facing our unit? Are they advancing towards us? Are they on higher ground? Are our flanks or rear threatened by the enemy? Are we under fire or is our formation disorganised for any reason, by having passed over a rough stretch of ground maybe? These points and others appropriate to the situation are added to form the minus side of the assessment, but we are not yet ready to make a comparison because we have yet to take into account the imponderable, that element of morale which cannot be predicted and which can at times convert the staunchest men into cravens, or can send a polyglot assembly of peasant levies into a headlong attack on a strong enemy.

To determine this uncertainty quotient, an excellent and prestigious phrase, we use dice, three of these essential cubes being thrown or rolled and their sum added to what remains after the minus factors have been subtracted from the plus ones. The final figure is compared with a pre-prepared chart which details the various consequences of morale assessments, and which can result in action entirely opposed to that which the general, i.e. the player, intended—as in the case of the flight of the Scythians whose shameful performance was described at the beginning of this chapter. This unpredictability is a considerable part of the attraction afforded by the war game and the facet which causes it to approach closely the actual art of war which was described by Clausewitz as being 'not an exact science'. In contrast to a mathematical game such as chess where the pawn or the rook will always move as the player wishes, one must in the war game cater for the occurrence of the unlikely or even the seemingly impossible. To quote another master of war, Napoleon was questioned as to his military system and replied "J'ai toujours fait mon thême à deux façons", which dictum loosely translated indicates that he always had some alternative up his sleeve in the event of the unforseen nullifying his primary strategy. Even he knew that nothing is certain and that the wise general should prepare for the unexpected and unpremeditated.

This is particularly so when we are concerned with troops like the uncertain Scythians but much less so with disciplined forces, and it is essential that when we get down to establishing the uncertainty factor it must provide for a wider variation for barbarian behaviour than for that of stable and reliable regulars. Given that the physical factors in a certain situation are similar for both types, one answer is to use two different sorts of dice. For barbarians the ordinary type numbered 1 to 6 is employed, so that when adding together the three dice being used to determine the uncertainty quotient we can get quite a prodigious range from a minimum of 3 to a maximum of 18. This is exactly right for the purpose of calculating the wide range of possible action of which the barbarian is capable. In contrast, for the regular we have other dice, described as 'average', which read not 1 to 6 but 2, 3, 3, 4, 4 and 5. Without going into a lot of calculation we can see quite easily that with the '1' and the '6' removed, and the two middle numbers duplicated, there will be a much greater chance of getting a middle-of-the-road result than with the 'wild' dice. In effect, unless extraordinary circumstances prevail, regulars will almost always get a result indicating that they do their duty and carry out orders. Rarely will they get carried away but similarly only infrequently will they fail to do their duty.

There will naturally be specific occasions when morale has to be tested. Many of these are self-evident—when a unit first comes under missile fire in a battle, when losses have been exceptionally heavy in a battle, or when the unit is about to charge or indeed is going to be charged. There is no special mystique about this system, for it prevails in varying ways in all sorts of war games in all periods, but it is particularly apt in the ancient era because of the special conditions found there.

Quite simply then, the sum of plus and minus circumstances together with the dice throw is compared with the reaction chart, where in most cases the directive will be found—'Obey orders' or something similar. It may be 'Stand fast' for a specified number of periods, indicating uncertainty or inability of the unit to screw its courage to the sticking place for an attack. At opposite ends will be 'Break and run' and 'Charge the enemy', the latter irrespective of orders of course. Either the break or the charge can be equally disastrous and once again the moral can be pointed from my own wargaming experience.

It was indeed a traumatic episode, and a splendid illustration of how a well-conceived plan was initially successful and how it was circumvented by the quite unexpected turn of events. The battle involved my army of Persians of about 400 BC and one of Germanic or Gaulish tribesmen organised and directed by their war chief, Andrew Green, a leading member of the local 'ancient' coven and better known to his familiars as 'Andra the Berserker'. The terrain, not a very complicated one, is depicted in diagram 1 and the position of the contending forces is shown after some five periods (it was a

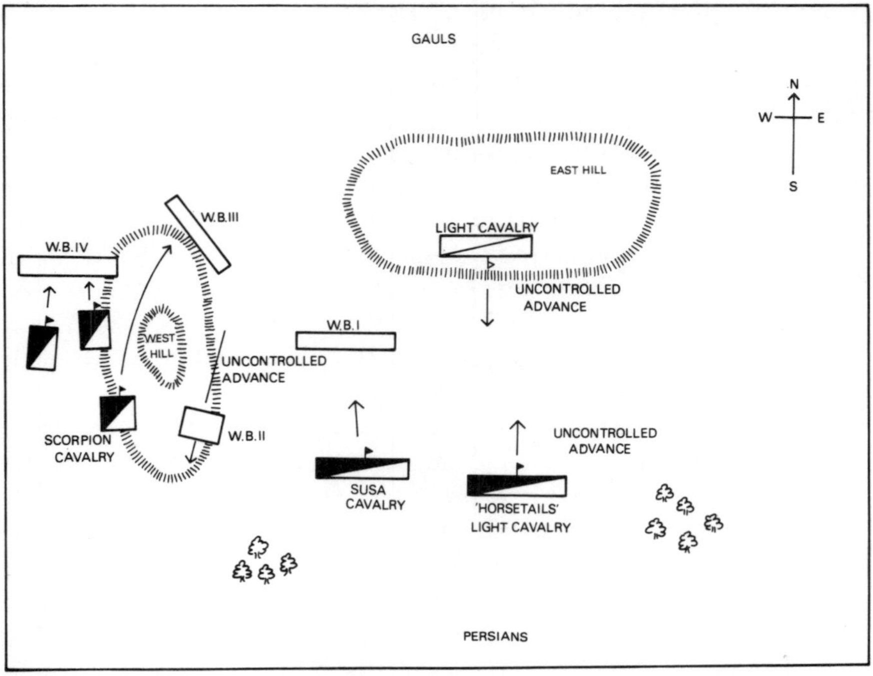

How a beautiful battle plan was circumvented by an uncontrolled advance by irregular cavalry and the author totally discomfited

simple encounter engagement with the opposing forces setting out from their respective base lines). The army of the dreaded Andra consisted of four tribal warbands armed with javelins, spears, throwing axes, etc and a strong body of light cavalry. As something of an experiment I fielded an entirely cavalry army, three regiments of light horse and two of heavy, the whole amounting to 110 figures. With his fewer cavalry, the enemy was numerically considerably stronger, but this disadvantage I made up for in mobility. In such a game players start with armies of the same points value and cavalry, 'costing' more than dismounted men, would naturally be outnumbered. All the enemy troops were barbarian, but my heavy cavalry was regular and my light barbarian.

First, I was able to concentrate two light cavalry regiments against Warband IV with total success, charging it with much élan—my variable throws were excellent, something of an overkill in fact, and this warband more or less disintegrated. On the other side of West Hill, when Warband II had its first morale test, its general had a prodigious dice throw which, together with the surrounding factors, was enough to send it into an all-out charge which left it sitting out in the blue at the south-eastern extremity of the hill (such a move without orders lasts two periods at the end of which no further action can be taken until fresh orders are delivered to the offending unit). Following up the success of the left-wing light cavalry, the 'Scorpion' heavy regiment moved round the west flank of the hill, as shown, charged and broke Warband III.

So far my plans were going very well but alas, a large-size spanner was about to be thrown into the works. On the right my aim was to move up the 'Horsetails' regiment of light cavalry and tempt the enemy horsemen on the hill to charge. If they did, the 'Horsetails' were to fall back, luring the enemy forward and enabling the 'Susa' heavy cavalry to swing round to make a flank attack, while the 'Horsetails' turned about and attacked the enemy front. At first it could not have gone better, for when the Germanic light cavalry perched upon the hill (East Hill) had its first morale test, it too went into the two-period uncontrolled charge. Magnificent, I thought, as I cast down the three 'wild' dice for the morale of the 'Horsetails', a barbarian unit, of course. Horror! The three dice produced a total of 14 which, together with the normal military factors, was enough to send *my* regiment into a similar headlong charge, quite scuppering my splendid plan. Instead of massing two regiments, 44 cavalry in all of whom 23 were 'heavy', against 30 enemy, it was the latter, with the additional impetus gained from a downhill charge who collided with my 'Horsetails', numbering 21. This at once put the focal point of the cavalry fight so far forward that to prevent Warband I getting on to the flank of the 'Horsetails' I had to charge it with the 'Susa' regiment (no trouble here with the regular unit's morale). At first it went well, but after a couple of periods of mêlèe my 'Horsetails' broke and fled, and then the 'Susa' people, unlucky enough to have taken on a warband of high-grade barbarians armed with the deadly throwing axe (the 'francisca') also decided that discretion was the better part of valour and similarly fled. My entire right was crushed, as indeed was that of enemy, but when time was called—there was a playing limit—it was obviously a Germanic victory, the loss of the 'Susa' heavy cavalry being decisive on a points comparison.

It was a salutory lesson, the moral being never to put one's implicit faith in barbarians, a precept I have endeavoured to adhere to since.

V
Egypt of the Pharaohs

In this survey of the art of warfare in ancient times (with particular reference to how its various aspects can be adapted to the war game), lack of space precludes any in-depth examination of the epoch in its entirety. There is the additional consideration that if these pages are to be of any practical value to the wargamer, it seems advantageous that we should concentrate where possible on those periods and nations whose armies are most viable for wargaming purposes. This of course assumes that the ordinary wargamer, unless he has some overriding interest in some special historical era or particular people, will not have much inclination towards an army which has little or no chance of success against whatever opponents he might be likely to meet. There is really no great satisfaction in being consistently defeated through the deployment of an inferior force. So although the number of warring nations and peoples in our period can be numbered by the dozen, many of them will be treated quite perfunctorily if in fact they are even mentioned, and wherever possible I shall dwell at greater length on those whose armies proved at least moderately effective while they existed, and which present-day wargaming experience demonstrates are forces to be reckoned with by any adversary who meets them upon a war games table. One must not necessarily feel, of course, that an enjoyable war game must inevitably be a victorious one—on the contrary, while one naturally and properly plays to win you may well find, as I have done on many an occasion, that some of the most stimulating and interesting battles are those which have ended with one on the losing side. Such an instance was the defeat I recounted in the previous chapter, where indeed I found the accurate simulation of barbarian morale most satisfying despite my discomfiture.

With this preamble over we can begin our study of ancient warfare in the area from which most phases of human activity seem to have stemmed—in the "Fertile Crescent" of the Middle East of which Mesopotamia roughly forms the centre. It is in this region, peopled by Akkadians, Sumerians and Elamites, that we find the origins of warfare, the situation where a group of people operating as a social or racial entity takes up arms against another such agglomeration. There is some evidence that as far back as the third millenium BC the people of a city named Sumer were a military force of some consequence in the valleys of the Tigris and the Euphrates. Like many other contemporary peoples, however, the Sumerians cannot be considered as being of great value or merit from our eminently practical and materialistic wargaming point of view. Their armies if reproduced on a war games table, stand little chance against more sophisticated and technically more highly developed troops. We therefore simply ignore them and move onwards in time as far as the neighbouring land of Egypt, whose kings—or Pharaohs—were really the first to field an army whose organisation and discipline compare very favourably with many we find at a much later date.

Now as we well know, Ancient Egypt had a political existence of thousands of years, but it was by no means a great military power during all these centuries. In this long span the country went through any number of political upheavals, revolutions and changes of dynasty, and was sometimes ruled by powerful and determined men, sometimes by effete and ineffectual Pharaohs. It might be a good idea then to choose one particular moment in the country's long and colourful history when, as far as its army was concerned, it was at an all-time peak of efficiency. This is exactly what we shall do for the benefit of the wargamer whose eye tends to light up at the sight of a miniature Egyptian army arrayed in battle order (an understandable phenomenon in any enthusiast who has seen *The Ten Commandments*, *Cleopatra* and other such cinematic epics). After all it is not unreasonable to assume that if a player favours a particular nation he will want to choose an army from it at the zenith of its military prowess.

Thus we can safely by-pass the earliest centuries of Ancient Egyptian history, even including the reigns of Tuthmosis III (1482-1450 BC) and Horemheb (1353-1319 BC), neither of whom were inconsiderable warriors, and light upon their successor of a couple of centuries later, Rameses II (1299-1232 BC), for it was under this very great leader that Egyptian military technique really reached its supremacy and it would be an Egyptian army of this period which could best be expected to give a good account of itself in any battle, even with armies of much later and possibly more technically advanced civilisations. Indeed it is fair to think of Rameses as being the very first of the Great Captains of history and in his organisational ability and his tactical knowledge he resembles Philip of Macedon, father of the mighty Alexander. It is true that in the years before Rameses came to the throne of Egypt much experience had been gained by the Egyptian armed forces in wars with neighbouring states, notably in Syria, and Rameses was able to use this to bring his soldiers to a very high degree of proficiency and to make the most of their military expertise. Not only did he develop the efficiency of his army, he enlarged it very considerably by increasing the three existing divisions of the army—Amun, Pre and Sutekh—by a fourth, the Ptah, and these were subsequently, by a process of doubling, converted to a type of army corps, each of two divisions; Rameses himself acted as commander-in-chief. It seems that, anticipating the days when Napoleon first organised self-contained corps of infantry, cavalry and guns able to act independently, these Egyptian corps were similar, including chariotry, infantry and ancillary troops.

Weaponry in this highly organised army was of surprisingly good quality, the Egyptians having been quick to adapt the weapons of enemies to their own use and much too had been learned from the Hyksos, who had invaded the country and indeed ruled it from 1646 to 1546 BC, when they had been expelled. Infantry weapons were the spear, the axe, the sword and the bow, possibly the most characteristic of these being the 'khepesh', a heavy bronze sword with a sickle-shaped blade, very thick and strong, with the cutting edge on the interior side. Axes of various types were in use. Originally a rather massive, unwieldy weapon, it had been made smaller and handier, had a bronze blade and could possibly have been used as a throwing weapon. Heavy maces—formidable weapons—were also carried by the foot soldiers at times. The single-stave self bow had been replaced by this time by the powerful composite bow, increasing fire power tremendously. Spears and javelins were in use as well, the former probably being of the short variety, not more than six or seven feet in length.

However, what one automatically thinks of as being the primary military characteristic of Ancient Egypt is certainly the war chariot which, in the absence of cavalry, was used in great numbers. Possibly the reason for this was the fact that the available horses were far too small and weak—little more than ponies it seems—and were not capable of carrying men into battle.

The chariot is certainly worth a few words of description, chiefly to explain the way in which it was used. Customarily drawn by two of the small horses mentioned, it was incredibly light, its construction being such that at the ultimate stage of its development, a single man could carry one on his shoulders with ease. This should explain one most important point concerning this vehicle—it could not in practice be used as a shock weapon, ie it could not be driven headlong into the ranks of an enemy to accomplish its task by weight and impetus in impact. One fundamental rule of warfare, and consequently of war-gaming, must be that in ordinary circumstances horses could not be made to charge home on unbroken and steady infantry, particularly if the latter were armed with long spears or some similar weapon (for example the bayonet-equipped musket of a vastly later period). Even heavy chariots could not be expected to do this and we have noted that most Egyptian chariots were extremely light, although some carried men armed with long thrusting spears with which to exploit any gaps in enemy ranks previously made by archery, into which the chariots could pass. When chariots were used in battle their preliminary tactics were those of a skirmishing force and involved driving rapidly up to the enemy and wheeling away, hurling javelins or delivering as heavy a fire of archery as they could while doing so. This would be kept up until the enemy were sufficiently shaken to be engaged by the infantry coming up in support.

Chariots and bow-armed infantry (many of the latter being Nubians from the south) were the mainsprings of the Egyptian army although Rameses II, after he had defeated an invasion of peoples from the Mediterranean (they were called Sherden or Shardanhu and could possibly have been from Sardinia) took a number into his service as mercenaries and they may even have acted as bodyguards. As might have been expected, in a formal battle the infantry formed the centre with the massed squadrons of chariots on the wings. The latter would make an initial advance to harass the enemy or to engage any opposing chariots. Of course, when chariots were used in such numbers they

Troops of the Pharaoh advance. Chariots, flanked by a column of infantry and preceded by javelinmen, with archers providing covering fire from the hill

became somewhat vulnerable, despite the skill and experience of the drivers, and it requires little imagination to envisage what could happen if but one horse was killed or wounded, or if a chariot overturned when one of the fragile wheels broke or buckled. The resulting chaos would be quite appalling. This of course, as we shall see later, is the weak point of a chariot army. With cavalry, a fallen mount can be avoided by its fellows with a little luck but even with chariots as highly manoeuvrable as the Egyptian variety this was by no means such a feasible proposition.

While tolerably well armed, the Egyptian was less fortunate in the matter of armour and his defences were rather scanty, his clothing being limited to the customary loin cloth, armour of any kind being the prerogative of the officers and men of high rank, and possibly that of the chariot warriors who as time went on became almost a socially élite class. Most of the line infantry were provided with a wooden shield, rectangular in shape, rounded and rather wider at the upper end, and normally covered with oxhide, hairy side out. It was occasionally reinforced with a bronze rim. The Greek writer, Xenophon, recorded that very large, almost man-size shields were used by the Egyptian contingent in the army of Croesus, but this was much later than the period we are considering.

This then is a very brief description of what made up an Ancient Egyptian army. It may be thought a trifle suspect from the wargamer's point of view, but we shall see. My first ancient army was an Egyptian one, but I was subverted by Persians when these figures began to appear on the market although they too may have had their day (Greeks can be most spectacular!). Nevertheless, from the point of view of mere appearance the Egyptians have nothing to fear and if one spends a little time on the painting and decorating of one's chariots, and adds the picturesque paraphernalia of standards and ensignia, one can assemble, as our illustration shows, a most attractive force. Furthermore, the wargamer can always have his personal figure riding in the Royal Chariot and wearing the double crown of Upper and Lower Egypt!

VI
Assyria

As succeeding Dynasties produced rulers ever more incapable and inept, and the power and glory of Pharaonic Egypt waned, there arose in the north-east yet another great kingdom, Assyria, whose kings were speedily to fasten their grasp upon much of the Middle East. For many hundreds of years a tiny Assyrian state had existed until about the ninth century BC when its people began to push their country's borders further and further westwards from their homeland in the hilly regions about the upper reaches of the Tigris, north of ancient Babylon. Under two kings Ashurbanipal II (883 to 859 BC) and Shalmaneser III (858 to 824 BC) this territorial extension was carried out with energy and determination. It was the beginnings of a blatant policy of expansion and conquest which lasted some three centuries until the Assyrian Empire fell before the vigorous assaults of the combined Medes, Scythians and rebel Babylonians.

Militarily the Assyrians are a remarkable and supremely interesting race, but in some of their habits and mores they were not a particularly attractive people, and they are really the only nation figuring in our period—including the Romans—which used sheer, naked terror and atrocity as instruments of military policy and political expediency. Generally throughout the time we are considering, human life was in no circumstances rated very highly, and even the Greeks, whose name has come to be fairly synonymous with the ideas of culture and civilisation, were not entirely guiltless in this regard, and one reads frequently of massacres of prisoners and similar events during the lengthy wars which racked the city states of Greece. These however were the exception rather than the rule, but this is far from being the case with the Assyrians. Cruelty was deliberate and planned, a lynch-pin of policy, mass flaying and impalement being but two of their more spectacular methods of disposing of conquered peoples, while their estimates of enemy battle casualties must be considered as being of some accuracy as they simply 'counted heads', this being done quite literally, the said heads, summarily and violently detached from their bodies, being piled in heaps in front of the king of the particular time. Apart from this distressing tendency though, some of the Assyrian kings were truly of a larger than life character. To the two already named one can add Tiglath-Peleser II, Sargon and Sennecharib, all of whom took their armies on a dramatic career of conquest. Thanks to their complete dedication to and an utter preoccupation with warfare, we are able to view the entire gamut of their military activity so imperishably and vividly depicted in the magnificent friezes and bas-reliefs in the Assyrian Galleries of the British Museum, where preserved for posterity is practically every detail relating to the Assyrian armed forces and how they were used. They were without doubt the most highly organised race before the advent of the Romans.

By the time the Assyrians began their conquests, rather more powerful horses than those available to the Egyptians were apparently in use, the Assyrians being the first to use cavalry—that is, soldiers mounted on horses for operations in this role rather than using the animals as means of transport from one point to another—mounted infantry, in other words. Assyrian cavalry vary in type, some are shown armed with spears (the stirrup has yet to make an appearance and the spear would be used for a free hand thrust rather than couched under the arm in the manner of the medieval lance), others with the bow. This weapon in fact was used by a high proportion of Assyrian soldiery, both horse and foot. The chariot however was for long the main strength of the army and during almost four centuries of continuous warfare it underwent many modifications, sometimes being drawn by two horses, sometimes by four, and with the crew similarly varying between two and four men, the fighting element of the crew being armed with spears or bows to provide both mêlée and missile power.

Again, as we did with the Egyptians, it seems a good plan for war-gaming purposes to select one period for mustering an Assyrian army. The choice is a wider one than for the previous nation, the different periods providing numerous armies of varying composition so that both the player who favours chariots and the one who prefers a high proportion of cavalry (which incidentally was used in increasing numbers as time went on) can make a selection accordingly. Following the precept that has been proved in real life, in the ancient war game the chariot is at something of a disadvantage vis-à-vis cavalry. The latter is more flexible in use and less subject to disorganisation, as well as being much more capable of coping with rough or difficult terrain. History shows that chariots required something like a totally flat and even surface on which to manoeuvre. (Rather later than our period, at the Battle of Arbela in 331 BC the Persian King, Darius, actually caused the ground in front of his battle line to be levelled to allow his chariots to operate successfully—a considerable task which, in the event, availed him nothing.) Imagine a large body of chariots—either four-horse or two-horse types—and also a formation of cavalry occupying an equivalent frontage. Obviously, any orderly movement will be, if not the easiest thing for cavalry, very much more difficult for the chariots. Should a simple move such as a wheel to the right or to the left have to be executed (one flank standing fast as pivot, the other moving round in the required direction), it will be abundantly clear that with a large number of chariots, each with two or more horses to be controlled, great care must be taken to ensure that wheels do not collide or interlock, and therefore the movement would have to be carried out much more slowly by chariots than by cavalry. Like many other essential factors, this will be reflected in any good set of war game rules.

So, we shall choose a point in Assyrian history where, although the chariots play their inevitable and important role, they are not entirely the dominant arm, and where there is a considerable leavening of cavalry, whose mobility and manoeuvrability compensates for the deficiencies in the chariot arm. We select then the Assyria of circa 700 BC, shortly after Sennecharib had succeeded to the throne, a time when the Assyrian military machine was at its peak.

It is obviously not possible to examine every detail of the Assyrian army nor to consider why it should have been so successful, other than to note that the entire civil administration and national economy were completely geared for war, many innovations—such as conscription—being repeated in the organisation of many later powers. Incidentally, one most important factor was that iron was now in general use for both armour and weapons, the superiority of this metal over bronze being obvious. Scientific siege craft was developed to a very high degree by the Assyrian army, as many of the nations which offered resistance to the conquerors realised that they would have a poor chance of success in the field against such a foe, and consequently shut themselves up in walled cities, having to be winkled out by means of a formal siege with towers, battering rams and other engines.

We shall consider the field army of Sennecharib, whose reign extended from 704 to 681 BC. Let us take first the infantry, not as yet the 'Queen of battles' as it was to become in later years, but nevertheless of great importance in the military scheme of things in Assyria. The spear-armed portion of the arm was equipped with an iron helmet ornamented with a crest (from which probably the later Grecian helmet was derived), and wore a type of fringed kilt over which was a shirt of mail, this consisting of very small iron plates stitched to a soft leather base. At one time it was thought that this protective garment was built up of layers of glued flax but it seems to me more likely that the iron mail is correct—there was plenty of this metal available at this time and certainly in 500 BC small rectangular iron plates were used for body armour in Egypt. (There is, in fact, such a mail-shirt in the British Museum.) Round, very 'dished' shields were common equipment, plus a stabbing spear and a short, straight sword. The spearmen possibly made up a fairly élite body in that their role was of a shock nature and certainly their training was extensive and prolonged.

Probably even more numerous than the spearmen were the archers, equipped with the composite bow, possibly the favourite weapon of the Assyrians. They originally it seems wore an ankle length mail robe but this was shortened before our era, during which it was mid-thigh length or even shorter. For sieges special corps of shield bearers were employed. They carried huge, man-sized shields to allow the bowmen to concentrate on their immediate task without running much risk

Two of the potent arms of the Assyrian army; heavy chariots and armoured, spear-armed infantry

from retaliatory fire. In addition to these heavy archers, we find lighter types able to skirmish in open order. Slingers too were not unknown, their high-angle volleys being particularly useful when attacking a foe on high ground or behind city walls. The Assyrian galleries of the British Museum provide most vivid illustrations of these fierce, bearded warriors and show dramatically how much they were to the fore in action.

Cavalry of all types were extensively used by the Assyrians; these are the light variety

Cavalry followed much the same pattern as did the infantry, there being both spear-armed and bow-armed varieties, the former wearing armoured shirts, although they were not always equipped with shields. Weapons were the sword and a long spear, for which the bow was substituted in the case of the horse archer. As a point of interest there seems to have been a time when one horsed archer was accompanied by another rider carrying a shield to protect his comrade, but this would seem to have been a most impractical undertaking and I cannot think that the practice was of long duration. It is extremely difficult to imagine this taking place with archer and shield bearer riding at speed.

And so to the chariots. In previous reigns there had been a great deal of alteration in design, weight and so forth—the two-horse chariot was changed to a four-horse type with attendant increases in the size of the vehicle and in the number of its crew, but by the time of Sennecharib it appears that a mixture of the two was employed—

some were the four-horse/three-man type and the others having the lighter two-horse/two-man arrangement. The latter, which formed the more considerable part of the chariot corps, will occupy our attention. Again the crew was basically driver and archer, the latter occasionally doubling as spearman or javelinman. Sometimes for protection, the number was increased to three men by the addition of one acting only as a shield bearer, but three men in the confined space of a two-horse chariot would be something of a crowd, each tending to get in the way of the other. Like all units of the Assyrian army the chariot corps was disciplined to a very high degree and operated in units of 50, termed a 'kisri', each having its own standard. It is indicative of the military advancement of the Assyrians that there seems to have been a special kind of uniform for men in a unit and their insignia and shield designs were the same for all in a particular kisri.

Finally, as might be expected, there was a guard unit of spearmen —known as 'Quradu', the 'strong ones'—whose duty was to protect the person of the king. They were doubtless a very tough bunch, with admission to their ranks open only to seasoned campaigners, this providing a fine opportunity for the wargamer to raise his own elite unit and, by lavishing some special care on the painting of the figures, to cut something of a dash on the war games table. The experience of handling a regiment of well-painted, well-produced miniatures is a most rewarding one. The trouble is that few of these guard units ever seem to be able to live up to the wargamer's expectations when it comes to the crunch.

All in all the Assyrian army was a most formidable opponent and existing statuary and reliefs show them operating in the most diverse

More Assyrian warriors; heavy infantry, with knee-length mail shirt, covered by javelinmen

conditions. They can be seen fighting in wooded and rugged hills and at the other end of the scale infantry are 'swimming' across rivers on inflated animal skins—about as early a commando operation as one can imagine. Thus, a wargamer who wishes such an army of light and heavy cavalry, with heavy and light infantry—bowmen, spearmen and slingers—plus an appropriate number of chariots, is very likely if he can make use of all arms successfully, to be a very worthy opponent to meet across a war games table.

We shall see the event.

VII
The Battle of the Apocryphal Well

"It came to pass that, in the reign of King Pelmaneser XXI of Assyria, he decided to wage war upon the people of Egypt, and to that end he sent an army across the Waste of Garan, not knowing, as he did so, that Horemhabib of Egypt had likewise concluded that conflict with Assyria would be to his great advantage and had sent forth his own army to do battle with the host of Assur."

(From *The Pelmaneser MSS*)

This is the background to the battle whose progress we are about to relate and which concerns the fortunes of the two advance guards—the one Assyrian, the other Egyptian—as they were resolved by a spirited encounter which took place under my aegis on my war games table and which, it is hoped, will in some small measure provide us with our first experience of an actual ancient war game as it occurred.

In military terms and in slightly more detail, the scheme demanded that the advanced guards of the two armies should march rapidly ahead of their respective main bodies deep into the Waste of Garan, and secure the possession of a most important well which, quite fortuitously, lay in the very centre of the desert, without the benefit of whose water the troops in such an inhospitable environment would suffer very considerably from thirst, with a consequent lowering of their morale as well as their fighting power. The actual well and its immediate surroundings are shown in the maps, where it can also be seen that it lay in an area encompassed by three low hills, whilst other hills lay further away to east and west, together with occasional clumps of trees. The entire country was rough and uneven, not indeed sandy desert but barren and undeniably uninviting. It need hardly be

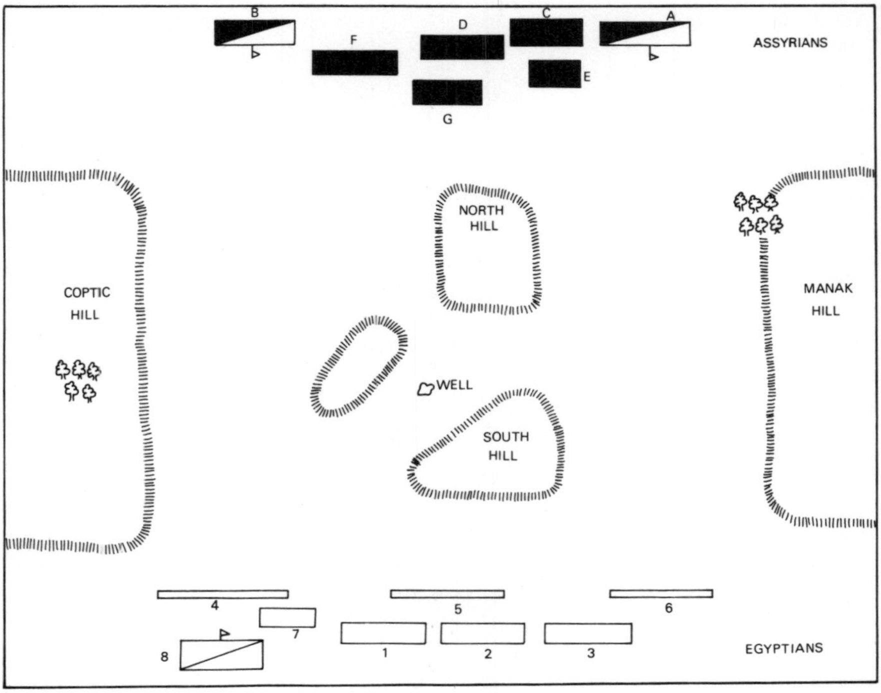

Map 1: the initial situation

Assyrians	*Egyptians*
A—Eshnunna Heavy Cavalry	1—Ptah Infantry
B—Zarzi Light Cavalry	2—Ra Infantry
C—Tutub Infantry	3—Senekht Infantry
D—Akalate Archers	4—Kush Archers
E—Hassuna Archers	5—Koth Archers
F—Tarbisa Infantry	6—Ka Archers
G—Repiquum Light Infantry	7—Lacish Javelins
	8—Chariots

pointed out that there were no roads. The orders for both forces were identical and simple—in effect that each should advance and seize the well, preventing the enemy from doing likewise, and await the arrival of the main army. This was due in a day's time and it was assumed that the two advanced guards had arrived on the war games table late on the previous day, had lain upon their arms all night and with dawn were deployed as shown in map 1, ready to begin operations. One of the various mechanical details of the battle is that we allocate 12 periods of fighting to a single day. This may seem odd, but it can be explained easily. There is always—and this is well known—a tremendous amount of hanging about in warfare, and if we take one move to represent an hour it will be occupied by 50 to 55 minutes of idleness and the remainder of the time by violent action. As to the particular war games table in use, it measures 9 feet by 7 feet, a goodly size by any standards.

With the sun about to rise over a stage more or less set, a word about the two contending armies is necessary. Their points value (or fighting value, that is) were equal, and the Assyrians, whose chariots had been left behind with their main army, mustered the following regiments:

Eshnunna Heavy Cavalry (armour, long spear and shield)	20 figures
Zarzi Light Cavalry (shield, short spear)	20
Tutub infantry (full mail, shield, short spear, bow)	26
Akalate Archers (bow, sword)	30
Hassuna Archers (bow, sword)	31
Tarbisa Infantry (half armour, shield, spear)	24
Repiquum Light Infantry (javelin, shield)	27
A total of	178 figures

To meet the Assyrian horde, the Egyptians mustered the following:

Ptah Infantry (spear, shield)	45 figures
Ra Infantry (spear, shield)	43
Senekht Infantry (spear, shield)	43
Kush Light Archers (bow, sword)	22
Koth Light Archers (bow, sword)	21
Ka Light Archers (bow, sword)	21
Lacish Light Javelins (javelin, shield)	18
A total of	213 figures

Plus the Pharaoh's Chariot Squadron—eight chariots, each with driver and one-man crew, the latter having both bow and javelins.

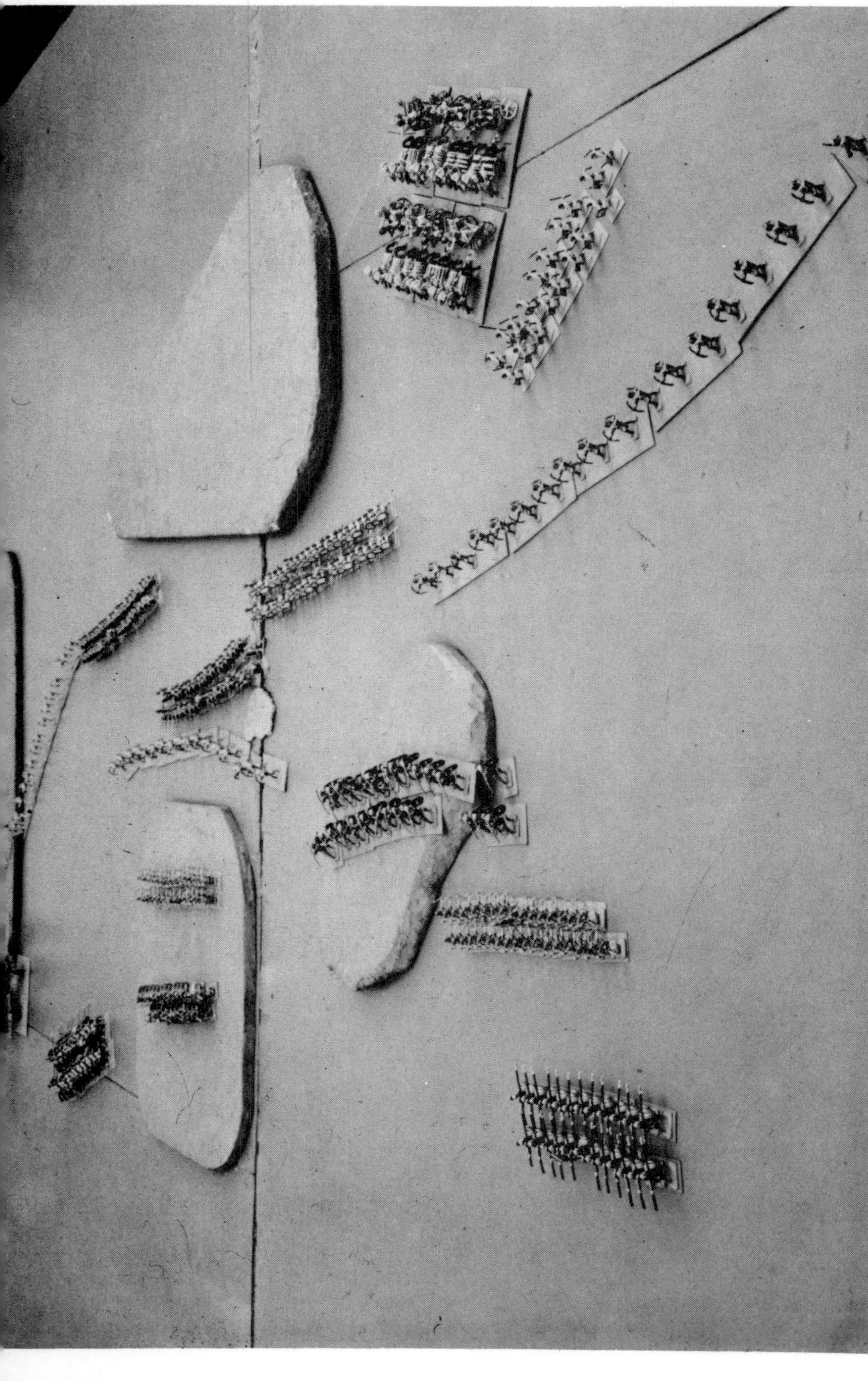

The Battle for the Apocryphal Well: the situation on period 5. The Egyptians have passed the Well and everywhere bringing the the Assyrians under archery fire

If we refer back to our suggestion (in chapter II) that it is a convenient ratio to make one war game figure equal 20 actual men, then we find that the Assyrians numbered 3,560 fighting men and the Egyptians 4,260 (exclusive of chariots)—quite appropriate strengths for their purpose.

It is evident that numerically—actually as well as representationally—the Egyptians were stronger, but the weight of the Assyrians amply made up for this deficiency—particularly when we consider the Tutub Infantry, with full mail protection and dual weaponry, and the Eshnunna Heavy Cavalry.

A final point before battle is joined concerns the personalities taking part. Two generals commanded on each side and they, as events proved, acted most amicably together. On the Egyptian side, Ian Osborn and Andrew Green split the command, and in common with their opponents both were quite unused to operating with the troops they actually had, the former normally being a Macedonian in the guise of Alexander the Great, while the identity of his colleague as Andra the Berserker has already been referred to in these pages. Facing them as joint Assyrian generals were Derek Casey and David Matthews, who had this in common that when their own armies took the field, the war games table groaned beneath the iron tread of Roman legionaries and Praetorian Guardsmen.

And so, with troops and generals in presence, not quite eyeball to eyeball but near enough, the die was cast, or in other words, the first move was made. The immediate aim being to take possession of the well, as laid down by their orders, both armies moved directly forward without more ado, the mass of medium infantry in the Egyptian centre covered by the Archers of Koth while the other light infantry hovered about the flanks and the chariots lingered prudently to the left rear. (It was later claimed that this was an astute manoeuvre designed by the Egyptian command to maintain a powerful reserve, but at the time uncharitable comments that the Egyptians were incapable of knowing how to use these vehicles were distinctly audible). On the Assyrian side there was an immediate extension of the army's front to right and left, cavalry swinging out in both directions, the Zarzi regiment apparently embarking on a wide turning movement round the west side of the wood on Coptic Hill. By move 3 archers were in action on both sides, the Hassuna Regiment taking heavy punishment from the Kothic bowmen. The former, a medium regiment with the men in close formation offered a far more vulnerable target than that provided by the light, open order Egyptians, and lost, during the third and fourth periods, no less than seven actual figures to their opponent's four. The initial morale tests were satisfactory for both armies, despite the high proportion of non-regular troops in the Egyptian ranks—all the archers and javelinmen, in fact. It is always something of an embarrassment for a general if a barbarian unit of light archers gets carried away by battle ardour and goes tearing off on a wild charge towards a more powerful enemy.

By the third period also, the Ka Archers ensconced on Manak Hill were causing no little trouble to the Eshnunna Cavalry regiment, now extended across the gap between North Hill and Manak Wood. (The word 'wood' is used in default of one more suitable as, after all, it was little more than a patch of scrub). There the Eshnunna remained—it seemed at this stage that the Assyrian commanders were at something of a loss just what to do with them, particularly as the mass of Egyptian infantry had passed South Hill, and indeed the all important well, and was now heading towards North Hill, which by this time was

Map 2: the situation, period 5

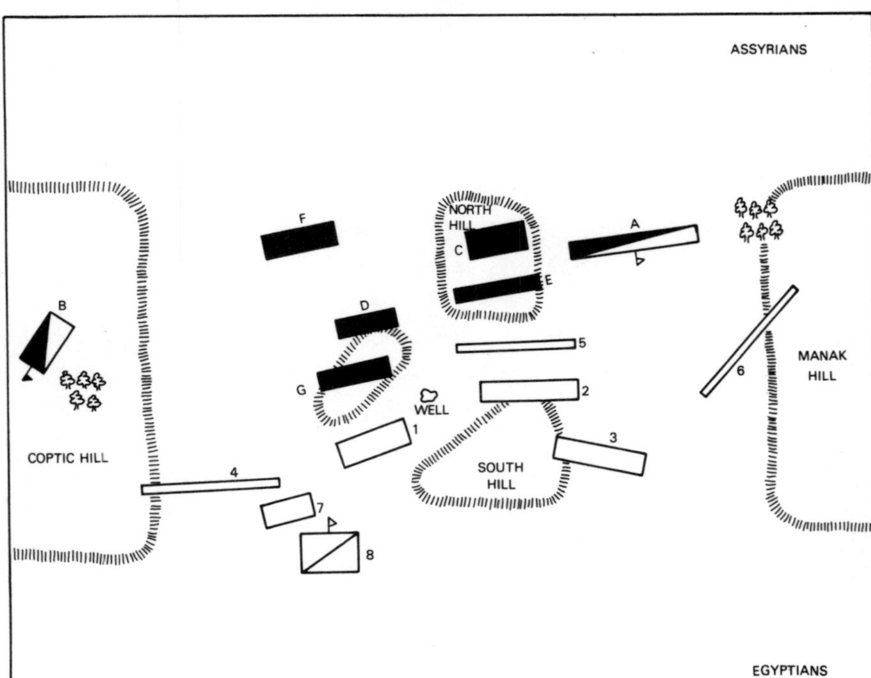

crowned by the Tutub Infantry supported by the Hassuna people who had already taken very considerable stick from the Kothic Archers.

At this point the Assyrians, taking a somewhat apprehensive look at the Egyptians en route for North Hill, brought up the Repiquum regiment of javelinmen as support for the troops already there and, on period 5, the situation was as depicted in map 2, where we see that the Egyptians have completely overrun the well and are now about to engage the enemy at close quarters. It was during period 6 in fact that the Repiquum regiment took the brunt of the assault by the Ptah Regiment, and despite having the advantage of fighting downhill, they were forced back (the 'push back' of 3 inches is the stipulated distance in such an event) and resulted from Ptah's inflicting 75 casualties on the javelinmen for the loss of only 40—in fact it was very close to an immediate 'break'—the 75 and 40 of course being representational casualties, 20 being one actual figure.

Following the success of Ptah, Ra now engaged the Tutub Regiment and also drove it back. In view of the somewhat parlous state of things on North Hill, the Assyrians now urged forward their light cavalry and at the same time the Eshnunna cavalry charged and broke the Ka archers who had been galling them for several periods and had indeed inflicted upon them the loss of six figures, a very high proportion of the total strength of 20.

Meantime, the Egyptian left wing bowmen, the Kush regiment, had pushed forward, shooting with some success at the Tarbisa spearmen, while themselves supported by the small javelin unit, the Lacish regiment. Worse was to follow for the Assyrians—in the struggle on North Hill, the Tutub infantry broke during period 9 (suffering more than twice the casualties it had inflicted) and fled, being seriously manhandled as it departed, and being followed very quickly by the men of the Repiquum regiment.

The end was indeed in sight. The Assyrian light cavalry came roaring round the wood and went forward in a headlong charge against the archers of Koth who had now faced south-west, being supported by the unengaged Senekht infantry. Much to their credit, the bowmen stood fast, but fortunately for them the Assyrian light cavalry were smitten hip and thigh by the Egyptian chariots which, stirred into action, delivered a devastating flank charge quite shattering the Zarzi regiment which collapsed in ruins. On period 10, three

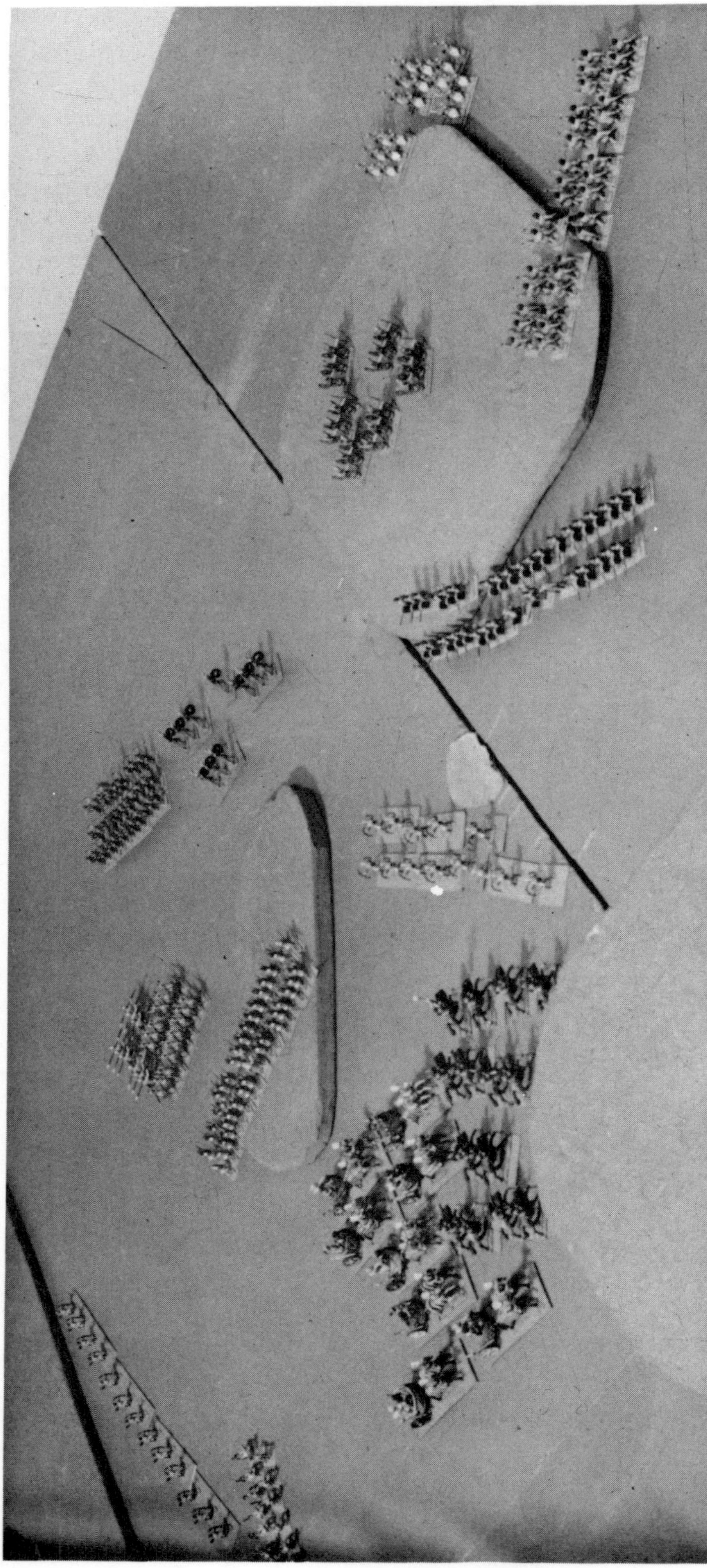

The Battle for the Apocryphal Well: the final stages. The Assyrians are in rout and in a final charge their light cavalry are taken in flank and rear by Egyptian chariots

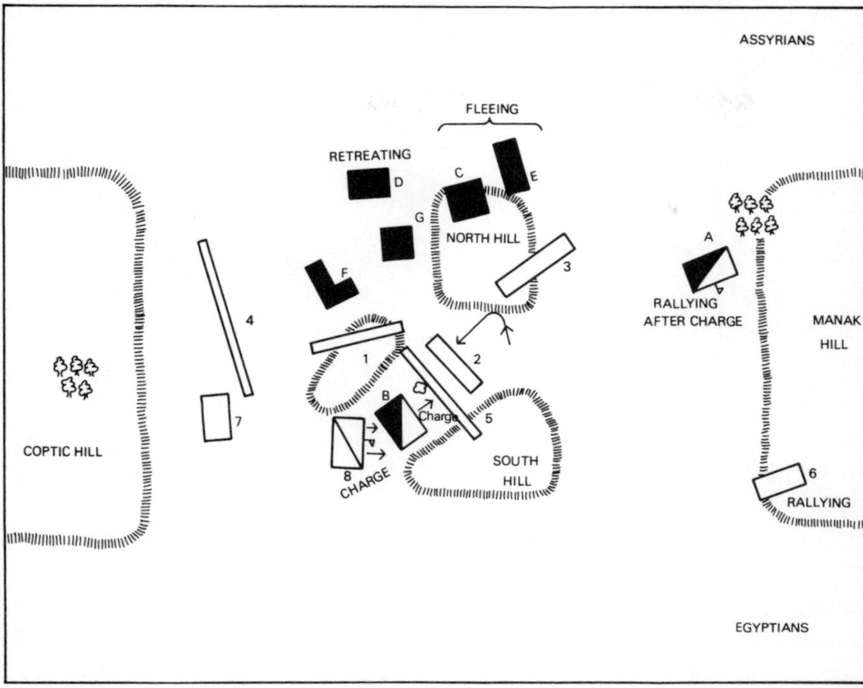

Map 3: final situation

of the eight Assyrian regiments were in flight and two others retreating—including Hassuna after its charge on Ka. The well was most effectively held by the Egyptians and it was abundantly clear that the Assyrian vanguard, or what was left of it, would have to fall back and rejoin the main body, prior to a major strategical re-think.

Materially, the victory was a decisive one for the erstwhile Roman generals, who in actual figures had lost but 38 to the Assyrian's 58, although to the latter total can be added the fleeing regiments who, should the result have been decided by a tally of viable points, would not have counted, being in flight.

Although the Battle of the Apocryphal Well provided nothing historically or militarily remarkable, it was a splendid little action, well illustrating a simple encounter battle. It occupied some three hours, although with a single player to each side it would have been rather longer. Time was saved to some extent by my officiating as clerical officer keeping totals of casualties, etc. Like every war game in which I have participated, it was replete with fun and excitement.

VIII
The Achaemenid Persians

Now that we have tasted a little action we can proceed without delay with our study of ancient warfare, and next we have to consider the power that arose from the ruins of Assyria. The last Assyrian monarch of any consequence, Ashurbanipal, died in 626 BC, and following his death numerous small wars arose between a variety of factions whose struggles for supremacy continued for several years. A unity was ultimately established which finally wrote 'finis' to the long career of Assyria, an event which can more or less be dated to 612 BC, when Nineveh fell. From the welter of contending forces there emerged the Persian Cyrus the Great who, with his son Cambyses, conquered the territories of the former Assyrian empire as far as and including Egypt, together with much of what we now know as Asiatic Turkey. The policy of these kings of Persia was one of aggression and territorial aggrandisement and it was in pursuit of this aim as it affected the west that they collided with Greece. From this conflict springs much that is of historical interest and of the greatest fascination in so far as it concerns the war game.

The first thing to be realised about these kings of the Achaemenids (this was the family name and is specified in this chapter because Persia had other rulers at a later date) is that in many respects they brought considerable advancements to many aspects of civilisation in their sprawling empire, which in its heyday stretched from the shores of the Aegean Sea to the borders of India. Although complete despots without the slightest regard for human life when it interfered with their plans, they are not wholly to be thought of as blood-thirsty monsters. The civil administration they created, although possessing a normal amount of corruption and venality, was nevertheless of marked efficiency. Indeed in the sphere of communications, the Persian road system and postal service were far ahead of anything Europe was capable of producing for many centuries.

Militarily, of course, much was inherited from the Assyrians, although Persia suffered from the great extension of her borders which forced the authorities into including in the army great numbers of levies and troops whose fighting qualities were not always of the highest and whose enthusiasm for the Achaemenid monarch was negligible. Politically too the country suffered from its division into governments—'satrapies'—which in some cases were practically autonomous, certain of the satraps, who usually came from the ranks of the Persian aristocracy, achieving a near independence. Numerical strength seems to have been of some consideration in the eyes of Persian generals, and although one must take with a very large pinch of salt the huge armies many ancient historians allege were gathered by the Persians for their attacks on Greece and for their defence against Alexander the Great, certainly the armies assembled by imperial decree must have been very strong, as far as mere numbers were concerned.

Alas though, Persian military thought, if not actually stagnant, was far from progressive and during the two centuries or so during which the Achaemenid Empire flourished only a very little forward movement can be discerned, the army of Darius III in 331 BC being not greatly different from that of the first of his line, Cyrus the Great in 529 BC. Thus, when we come to choose a Persian army (which as I have learned myself is not without its benefits for the wargamer) we find that it does not much matter under which of the Achaemenids we are to serve.

Naturally, as in nearly all societies where the dominant factor in matters secular is a strong aristocracy, we find that the most important arm of the fighting forces of Persia was the cavalry—in complete contrast (as we shall see later) to Greece, where the citizen soldier of the democracies was an infantryman. This Persian aristocracy provided a splendid body of mounted fighting men who, together with the semi-regular tribes from the northern portions of the kingdom, formed the only section of the Persian host which was of much account. It is fitting then that we deal with the cavalry first.

A large variety of horsemen were to be found, ranging from the heaviest to the lightest and, due to the old Assyrian influence, the composite bow was their favourite weapon. As time went on however the bow seems to have lost the favour of certainly the light cavalry and, despite the success of mounted archers in various battles, even against such doughty opponents as the Greeks, its place was taken by the javelin, two such weapons being carried by each cavalryman. As the shield was apparently not favoured by the Persian light cavalry they

Cavalry was a most powerful component of the Achaemenid Persian army; here light horse charge down a hillside

were extremely vulnerable to a more strongly armed and better protected enemy who would consequently be eager to come to close quarters. It is understandable that a man with a bow will prefer to operate from a safe distance and generally be reluctant to become involved in hand-to-hand combat. Some of the best of the light cavalry incidentally were the Scythians, although these could just as readily appear in the ranks of an enemy, they also being particularly fine archers who had a small axe or dagger as additional armament. At the other end of the scale we come to the heavy cavalry. Naturally not so numerous, they were usually equipped with iron or bronze scale armour shirts which fell anywhere between the waist and just above the knee. They could also be armed with javelins—certainly in the early years of Persian ascendancy—although these later gave way to thrusting spears. In connection with the latter we are still speaking of the pre-stirrup era and the spear is used, not as a lance, but rather as a free-hand stabbing weapon. This may be one reason for the fact that Persian cavalry never appear to have charged in knee-to-knee formation but rather in the manner described many centuries later by the French as *en fourrageurs*—every man for himself and riding with little or no regard for the maintaining of any formation.

Of the heavy cavalry there was a portion which might be described as 'extra-heavy', the horse being armoured as well as the rider, the equine protection consisting of iron, bronze or leather scales on a cloth housing, this covering the horse's body in front of the rider's legs, as well as the animal's neck and head. Such a body of horse—generally named 'cataphracts'—is said to have been provided by the Scythians for the army of Darius at Arbela, the last Persian battle against Alexander the Great. Probably it would have been mixed in quality; the Persian cavalry was indeed formidable but lacked a first-class field commander to make it really effective.

The infantry of the Persian Empire was possibly even more heterogeneous than was the cavalry. Ancient Greek historians, to whom we owe much of our knowledge of the time, described in their works some of the exotic soldiery gathered together or conscripted by the Persian Kings for various campaigns they undertook. These outlandish peoples ranged from Arabs to Egyptians to Ethiopians, but the real hard core of the infantry consisted of Persians and Medes, the dominant and most important races of the Empire. As by far the greatest proportion of the army consisted of archers, it would seem that relatively few soldiers would be provided with any sort of armour, and if it were used at all it must have been worn under the voluminous robes worn by many of these troops. Various frescoes discovered in the ruins of Persian palaces and similar places show infantrymen in considerable detail but no armour is visible. The range of the bow—a composite one—was in the neighbourhood of 200 yards. Throughout the army native heavy infantry seems to have been entirely lacking, although in later years this deficiency was made up by the hiring of Greek mercenaries.

The best known, of course, of the troops in the Persian Army are the Immortals, a body which was founded by the first Darius, but which no longer appeared in the order of battle by the time of Arbela. They numbered 10,000 men and owed their title to the fact that there were always reserves available to fill up gaps due to battle casualties or sickness and the strength consequently never fell below the establishment. Another apparently rather specialist corps were the Cardaces. These men were a cross between light and medium infantry, capable, it would seem, of acting as skirmishers or alternatively of taking their place in the line of battle, as they did at Issus. Like the bulk of Persian infantry they wore a felt or stiff cloth head-dress with long lappets hanging in front of the ears, carried a small, circular shield and were armed with a short, stabbing spear. They were recruited from the Persian upper classes, which would indicate that while possibly not of elite calibre they were nevertheless of above average class. A few slingers were also to be found in the Persian army, who would be clothed and accoutred as were the more numerous archers.

As time passed it became increasingly clear to the Persian kings that their infantry did not have the ability to match the power and weaponry of their Greek enemies and they began to employ considerable numbers of Greek mercenary soldiers, of whom there were great numbers available about the middle of the fourth century BC and in Darius' last battle, which marked the end of the Achaemenids, the only infantry in his army were some 2000 Greek hoplites and the same number of his own foot guard. These hirelings were not automatically unreliable; those who fought at Arbela refused to change sides after the defeat and remained loyal to their employer until his murder some time after the battle.

Lastly we have the chariots. These had been employed at the Battle of Cunaxa in 401 BC but fell into disuse until Darius mustered some 200 of them for the Battle of Arbela. The drivers, however, had not received adequate training and, despite the levelling of the ground already mentioned, they were of little use and were speedily and effectively dealt with by the Macedonians.

This then has been a very cursory description of the army of Achaemenid Persia. For a variety of reasons, it really never had the opportunity of showing its true worth. In the invasions of Greece it was on very unfamiliar ground, and both at Marathon (491 BC) and Plataea (479 BC) it was both on unsuitable terrain and commanded by leaders who, if not actually inept, were certainly not of the highest quality. Later, against the genius of Alexander and his superbly trained and organised army, it stood little chance of success. Still, for the wargamer a Persian army has much to offer, and for the variety of colourful costumes apart from anything else it stands in a class by itself. Furthermore it really can provide the basis of a worthwhile fighting force, and in the hands of a wargamer of some skill and experience it should have its share of success. After all, from such a wealth of troops it is not difficult to select a well-balanced army, with light, heavy or extra-heavy cavalry, infantry both medium and light, as well as different missile troops and chariots. If one wishes to dispense with the colourful Immortals and opt for the army of Darius III rather than for that of Xerxes, one can have mercenary Greek hoplites—none better when it comes to heavy infantry. Of course in the earlier period during the invasions of Greece, certain of the Greek states, notably Boeotia, 'Medized' as the saying went (to the Greeks the Persians were in fact Medes) and allied themselves with the invader. Indeed at Plataea a substantial part of the Persian army consisted of Greek hoplites who gave a good account of themselves. So, with a Persian army the wargamer is in the fortunate position of having the best of both worlds, and it is my own experience that an army of 479 BC can be counted upon to be a rather tough nut to crack on the war games table. It is only fair to point out, though, that a considerable proportion of such an army, particularly the light troops, would of necessity have to be considered non-regular or barbarian and would be liable to the vagaries of behaviour that we have already noticed.

IX
The Greeks

Before the seventh century BC or thereabouts, warfare as practised by the Greeks was a highly individualistic sort of pastime, bearing little resemblance to the military system which was to succeed it. In terms of time the period might be termed the Heroic Age and, in effect, it was exactly what the name implies. Arms in general were little better than rudimentary, only the wealthiest or most powerful being able to afford the bronze armour and weapons which were available in only small quantities, and battles usually consisted of two motley arrays of ill-equipped men being brought together. From their none too willing ranks there would emerge a king, a general or a chief of some kind, whose purpose it was to do battle with a similar individual from the opposing side. They would not be on foot but would be riding in chariots, vehicles of exceedingly light construction not unlike those of the Egyptians and consequently highly manoeuvrable. The 'hero' was armed with javelins and a sword, the traditional type of armour and crested, enclosed helmet which was such a characteristic of the later Greek warrior had yet to be introduced. He may well have worn leather armour and a cap of the same material reinforced with sections of boar tusk stitched on to it. The bow does not seem to have been used—it was never a favourite of the Greeks.

Anyway, each contestant would whip up his horses and drive them at full gallop towards his enemy, not with the idea of a violent contact but rather with the aim of passing close to him and hurling a javelin or two in his direction. It is apparent that the highest degree of control over the horses was required and certainly such a combat must have been spectacular in the extreme, with the chariots wheeling around, being dragged to sudden halts and racing about all over the place in clouds of dust. If the javelins were flung without effect it would appear that the warriors would dismount and continue the fight on foot with their swords, these weapons being of insufficient length easily to reach from one chariot to the other. This sort of heroic conflict was more or less de rigeur at the time of the Trojan War, the classic instance of such a fight being that between Achilles and Hector, so vividly described by Homer. Normally, it seems that all the attendant troops—for want of a better word to describe them—played, more often than not, the part of mere spectators and departed for home when their betters had settled whatever dispute was at issue.

As time wore on, however, the classical Greek system of the 'polis' or city state developed, largely as a result of the mountainous nature of the country where each community in its own none too accessible area formulated its own political structure. Technology and wealth also increased, resulting in the appearance of a substantial citizenry, men who were affluent enough to provide themselves with the bronze and later the iron weaponry which was becoming increasingly available as production methods improved and the price came within reach of this rapidly growing section of the cities' population. Initially in the sixth century BC there emerged the fighting man we ordinarily regard as the archetypal Greek warrior, the hoplite. He was initially a citizen whose social position and means were such as to ensure his station in the fighting forces of the state or city to which he belonged and he received his name from the characteristic large round shield—the 'hoplon'—which was an integral part of his armour. It was some three feet in diameter and consisted, not of solid metal, but of a thin sheet of bronze into the back of which was hammered boiled leather —or *cuir bouilli*—a shield of solid bronze naturally being far too ponderous for even the strongest to carry. By this time the hoplite was furnished with the classical Greek helmet which almost completely enclosed the wearer's head, leaving slots for the eyes, and was usually fitted with a tall horsehair crest. Leather or metal body armour was worn together with greaves to protect the lower legs, while the principal weapon was a heavy spear some nine feet in length. A short sword was carried as a secondary weapon but was of far less importance than the spear. Arrayed in a solid, close-ordered mass, the phalanx of hoplites, anything up to eight ranks deep, when presenting their spears to an enemy showed a bristling front of several layers of glittering points, and it was such a phalanx which was one of the most important tactical features of the interminable struggles between the Greek states themselves and their warfare with successive Persian kings during the latter's attempts to invade and conquer the Greek homeland.

The advance of a Greek phalanx with its row upon row of great, overlapping shields and its multitude of spears must have been a truly fearsome sight, although as a military formation it had many disadvantages, despite its seeming power. First of all there is the question of manoeuvrability. Various orders of 'dressing', ranging from intervals of $1\frac{1}{2}'$ to $3'$ per man, were employed by the phalanx but it can be easily seen that such a mass of men, with shields interlocking and with long spears in hand, had to be extremely careful to preserve proper formation and any variation in the pace of different sections of a phalanx would immediately throw it into the greatest disorder. Hence, its rate of movement was a rather deliberate one and even a simple change of direction, such as a wheel, had to be carried out slowly and with the greatest care. Within the phalanx itself there were drill movements such as doubling of files which could be carried out but an overall manoeuvre—other than an advance straight ahead—was something to be thought of at least twice. This feature of the phalanx—and with hoplites providing by far the greatest proportion of the majority of Greek armies, forming in many cases the entire field force—ensured that a hoplite battle had to be fought on the most even and level terrain, for should any irregularity be met the phalanx more or less ground to a halt or dissolved into confusion.

However, when two phalanxes did collide they would do so probably at a slow run or a trot. There would be an almighty crash as the two shield-walls met and this would be followed by a sort of heave to and fro—a kind of rugger scrum—with not too many casualties being inflicted, due to the shields and armour, until by reason possibly of a continuation of the pushing movement in one direction, or by a break in formation or a weakening of morale one scrum, in a manner of speaking would collapse—the men making it up would turn tail and

Infantry formed the backbone of the classic Greek army; here are units of Boeotian, Corinthian and Spartan hoplites

run, thus exposing their backs to the enemy. This was the moment when the heaviest casualties were inflicted, the long spears causing great mayhem, and most hoplite battles were characterised by heavy casualties on one side and relatively few on the other. True, the chase was rarely maintained for any length of time, pursuit to the bitter end not being part of Greek military philosophy. The winners would erect a commemorative trophy, gather what plunder they could, assemble any prisoners as potential slaves and then set off for home.

Hoplite battles are easily reproduced on the war games table, and once the two phalanxes of miniature soldiery have engaged, there are two conditions for the break to which we have referred. If during one period of fighting one phalanx has suffered more than twice the casualties it has inflicted, it breaks and in the next period the figures composing it are turned sharply about and moved off in flight. During this initial break the victorious phalanx is able to strike at the undefended backs of the routed without any let or hindrance. The second condition is that one phalanx, having suffered more casualties than it inflicted during three successive periods (without these at any time providing the two to one ratio which constitutes an immediate break) and having been pushed back a prescribed distance at the end of each of the three periods, will thereupon break and run, the same free strike being awarded to the victors as in the previous case.

Another point concerning the phalanx which will not have escaped notice is that when organised in a solid mass, its flanks and rear were very sensitive, not to say vulnerable, to any enemy attack. This did not have great significance when one hoplite formation met another in the very formalised sort of engagement to which the Greeks were initially prone, but as time went on and a greater variety of foes were encountered, problems arose, especially when the new enemies had any light troops with which to assail the tender spots of the phalanx. Something had to be done to counter such a threat, and a corps of light troops was created, far less heavily equipped than the hoplites, whose duty it was to attack the flanks of an enemy phalanx or to protect those of their own hoplites should this be required. More of this in a moment, however.

Yet another feature of the phalanx was its tendency to 'drift' to its right during an advance. This was the result of the not unnatural inclination of each man to edge to his right to protect his unshielded side, the movement being presumably initiated by the right hand man in the front rank who must have been acutely aware of his figuratively naked side and would understandably move to his right to get it away from the enemy. Thus it frequently happened that, when two phalanxes were closing with each other, the right wing of each overlapped the left wing of the other. In passing, this is a feature which can quite readily be covered in ancient war game rules by laying it down that, for every 6″ of forward movement, the phalanx must drift ½″ to its right, in effect a kind of modified knight's move as in chess. The wargamer, however, might choose to ignore this, deeming that a compensatory leftward motion would be ensured by the officers of the phalanx—the choice is a personal one for the individual player.

What is important, however, when operating a Greek army on the war games table is the use of light troops to support one's phalanx. It seems that lightly armed and very mobile people existed, to a limited extent, even in what we called the heroic age, but it was when the hoplites found themselves operating away from their cities, particularly in the northern parts of Greece such as Thrace that they encountered societies where the city state had not evolved with its attendant array of heavy infantry. Instead there were numerous tribes and villages where fighting men operated individually, were fast in action and in uneven or rugged country presented a most serious threat to the slow moving and unwieldy phalanx. So there evolved new classes of fighting men, a few being raised in the cities themselves but many more being recruited from the northern provinces.

There were two classes of such troops, only one—the 'psiloi'—being truly described as light. They were purely missile men, sometimes armed with light javelins and slings, but almost invariably having no protective shields. Better known are the 'peltasts', so called from their shields, the 'pelte'. They were really a sort of light/medium soldier, and although frequently operating at long range like the psiloi, they could, when called upon, fight hand to hand with considerable gusto and effect. Thrace was the traditional home of the peltasts and they can be seen depicted on many Greek vases in museums everywhere. The shape of the pelte is of some interest, sometimes being round and sometimes being of a crescent or half-moon shape. It was usually of wickerwork covered with animal skin. When used correctly peltasts could be most formidable, that is, in the right conditions or in combination with other troops. With hoplites they formed a very effective body of soldiery. Two examples will

Greek light troops both infantry and cavalry; neither archers nor cavalry were extensively used by them

suffice to show their prowess. During fighting on the Island of Sphacteria in 435 BC Athenian light infantry and hoplites in a well directed effort defeated and took prisoner a force of Spartan hoplites, who themselves had not the benefit of supporting light troops. Another notable occasion was in 300 BC when a celebrated mercenary general, one Iphicrates, destroyed a body of Spartans with his peltasts.

Incidentally, the defeat of any Spartans in these centuries was a very special thing, their reputation being very high. The economy, life style and philosophy of Sparta were purely military and in general the city was totally geared for war, its army being far and away the most disciplined and organised in the whole of Greece. Although their troops had not assisted the Athenians in their victory over the Persians at Marathon in 490 BC, the stand of 300 of their hoplites under King Leonidas at Thermopylae ten years subsequently has, although in itself a defeat, become a legend. The following year, 479 BC, however, saw the combined forces of Athens and Sparta totally defeat the Persians and their Greek allies at Plataea, the Lacaedemonians, as the Spartans called themselves, proving their worth many times over.

Traditionally then the Greek soldier of this period—sixth to fourth centuries BC—was the infantryman, generally a heavy (although he could be light) while Greek cavalry was virtually negligible, what did exist being of a very moderate quality. It was lightly armed and ill-trained, even where it was fairly numerous, such as in the plains of the mid-north, and the concept of cavalry/infantry cooperation rarely if ever occurred to any Greek general, despite what should have been learned by at least the Athenians during their disastrous expedition to Sicily in the years 415-413 BC. This lack of coordination was a general hallmark of Greek armies, which were also notable for lack of missile power, the bow being unpopular although the sling was more in evidence. This lack of missile troops was made dramatically evident at the Plataea battle already spoken of when there was but a handful of archers to counter the prodigious arrow storm of the massed Persian bowmen. It was fortunate, in fact, that the latter's supporting troops crowded up so closely behind the archers that when the Spartans charged the Persians simply could not escape and were speared in thousands by the hoplites.

Despite such strictures as I have passed on the Greek military system of this period, it was far from being utterly useless and my own belief is that, like the Persian, it suffered greatly from a lack of really good generals, who could use all arms to advantage, choose suitable terrain and fight in appropriate conditions. Furthermore it was plagued by a political system which ranged city against city in wasteful and pointless local wars, wherein new feuds were stirred up and old ones perpetuated. All this ensured that a united Greece was only a chimera at least until a really great general could weld its various elements into a homogeneous force. As we shall see, he was soon to make his appearance.

In war game terms the Greek army is not to be despised, although the player could suffer from lack of fire power, for we must accept that not more than ten per cent of such an army could, if we are to be historically correct, be armed with the bow, although some slingers could always be employed. Cavalry also would not be too much in evidence, although by choosing the latest possible army within the period, say about 300 BC, one could include useful Thessalian and Boeotian cavalry, or indeed opt for a Thessalian army in toto, with cavalry amounting to twenty per cent of the whole. Although light, these were not to be despised. From the point of view of appearance, though, there is much to be said for a Greek army, as the illustrations should demonstrate, the striking Corinthian helmet with its high crest being one of the most impressive articles of military headgear ever devised, while the great shield—the hoplon—could be decorated with infinite variety. I have a unit of Boeotians whose shield device is a dolphin—in black—always on a different coloured background for each man. The effect is most pleasing—indeed quite splendid.

X
The Battle of Plataea—The Background

It will not have escaped the attention of the discerning reader that the Battle for the Apocryphal Well had rather more than its share of the fictional (although as a war game it was none the worse for that) but nevertheless, it provided, it is hoped, something of an aperitif for the main course of the wargaming meal. This involves something more portentous than the ad hoc Assyrian/Egyptian affair which has been described. I hope I shall not be misunderstood if I refer to the next practical demonstration of a war game as being a little more serious than the previous one, for it is far from my intentions that we should lose any sense of proportion—or indeed of humour—by laboriously taking the reader through a tedious and dull sort of operation such as might cause the less committed to run a mile. What I propose to do is simply to stage another war game, this time against an actual historical background, where we can compare, among other things, war game conditions with those of a real life engagement, note how the rules in use reflect actual military operations and how the tactical skill of the players taking part compare with the actions and decisions of their own historical prototypes.

To refight an actual battle as a war game is something I have done many times and in a variety of periods. Consequently I can claim that this section of the war games hobby is without doubt one of the most enjoyable and satisfying. Apart from the excitement of the actual war game per se, there is the very considerable pleasure to be derived from the reading and research necessary to establish the facts concerning the chosen engagement, the details of troops and terrain, the manoeuvres carried out and so on. Sometimes, of course, with very large battles it will be found necessary to divide the action into sections and refight whatever occurred in each separate portion, later correlating the overall tactical situation to establish a result.

This can readily be done in later historical periods, when battles spread over a front of several miles, but in the period of our immediate interest it will generally be found that a largish table can cope with the majority of the better known engagements, providing that the ground scale can be satisfactorily reduced, together with the number of troops taking part. More of this later.

Since we have recently dealt with the armed forces and military institutions of both the Greeks and the Achaemenid Persians, we might with advantage choose for the closer study of our next battle one of the classic encounters which took place during the lengthy wars between the two nations, and of the several which come to mind, none is more suitable than that which marked the final expulsion of the Persians from Greece—the Battle of Plataea in 479 BC. Overshadowed in popular legend by the drama of Thermopylae during the previous year (much as the highly successful Charge of the Heavy Brigade at Balaclava was, thanks to the power of poetry, forgotten in favour of the catastrophe of the Light Brigade in the same battle), it was nevertheless a crucial point in European history, as important on land as Salamis was at sea, and it ensured the turning back of the Persian tide and the end of attempts to impose an Asiatic culture upon Europeans.

It might be as well to give a very brief account of events leading up to the battle, if only to refresh our knowledge of the strategical *raison d'être* of the fight, and we need go no further back in time than about a quarter of a century, to 500 BC in fact, when Darius I of Persia was appealed to by the Greek colonists for assistance in a revolt against their parent cities. From this ultimately stemmed the first invasion of Greece, which was decisively defeated by an Athenian army at Marathon in 490 BC. For some time thereafter the Persians were fully occupied at home, largely in the putting down of various revolts, but it was apparent to all thinking Greeks—not too many at this time—that the storm would soon break again upon their country and indeed it did in 480 BC when Xerxes, successor to Darius, assembled a vast, if somewhat polyglot army at Sardis in Asia Minor, marched northwards, crossed the Hellespont and plunged into Thrace and Thessaly. The invasion route into southern Greece led through the Pass of Thermopylae and it was there that the Persian host was confronted by a small Greek holding force, of which the hard core were the 300 Spartans of King Leonidas. After three days of fighting the position

Colourful units of the Persian army passing a Greek temple in Ionia with archers in the van, followed by chariots and a detachment of the renowned Immortals

was turned—a traitor revealing to Xerxes a hidden path through the mountains—and the Greeks were overwhelmed and destroyed. As Xerxes marched south, however, the great Persian fleet which supported and flanked his land forces was engaged and defeated by the united Greek navies at Salamis. Despite having occupied Athens, Xerxes was alarmed for his lines of communication and decided to leave for home. He marched off to the north and the bridges over the Hellespont with much of his army, but behind him he left a powerful force under his general Mardonius to hold northern Greece.

During the winter of 480-479 BC little of import took place, but with spring came various diplomatic démarches designed primarily by Mardonius to create dissention among the Greeks—never a difficult thing to accomplish. Indeed it must also be remembered that in the army of Mardonius were the contingents of a number of northern Greek cities which had thrown in their lot with the Persians. Once again Athens, which had been evacuated by the Persians, was re-occupied and at this point the Spartans, after some delay (they were always very careful to consider the pros and cons of a situation before committing themselves to any definite course of action), mustered their forces and marched northwards towards the Isthmus of Corinth, arriving there sometime in July and were joined there by the Athenian and other Greek contingents. Once assembled, the entire Greek army marched off towards the territory held by the enemy, finally reaching Mount Cithaeron. As they emerged through the mountain passes the Greeks could see, from the foothills on which they stood, the camp of the Persians, extending on both sides of a large stockade just beyond the River Asopus as it ran from west to east between the two armies. The stockade was a large one, something like a square mile in area apparently. The armies were now to all intents and purposes in presence and at this stage it is proper that we note as far as possible their relative strengths so that when we set up our troops on the war games table they will be historically correct as to number as well as position.

Thousands of words, indeed volumes, have been written about the forces engaged and it is virtually impossible to arrive at anything really definite. One thing is certain—that the Persians were numerically stronger than the Greeks, and much more so in cavalry and fire power, Pausanias having no mounted troops whatever. We have already noted the preponderance of bow-armed troops in the Persian army and the lack of enthusiasm for this weapon among the Greeks, but man for man the Persian infantryman was poor stuff in comparison with the Greek hoplite. Even so, the Greek elements of the Persian army were every bit as good as their more loyal countrymen. As to mere numbers, though, we can do no better than accept the totals as furnished by the authority most recently writing about the battle. In round figures the army of Pausanias numbered 40,000 men, while that of Mardonius was 50,000, of which latter figure some 10,000 were cavalrymen.

However, the numerical inferiority of the Greeks was amply compensated for by their superiority in weapons and armour—the long spear against the short, the bronze helmet against the felt head-dress, the heavy hoplon against the light wicker or wooden shield, and so on. Of course, in fire power the Persians completely outclassed the Greeks and in a strictly shooting battle would have carried the day, but this was not to be as we shall see.

Before engaging in the battle ourselves it is necessary to go through the action as it took place, and this we can do with the aid of map 1,

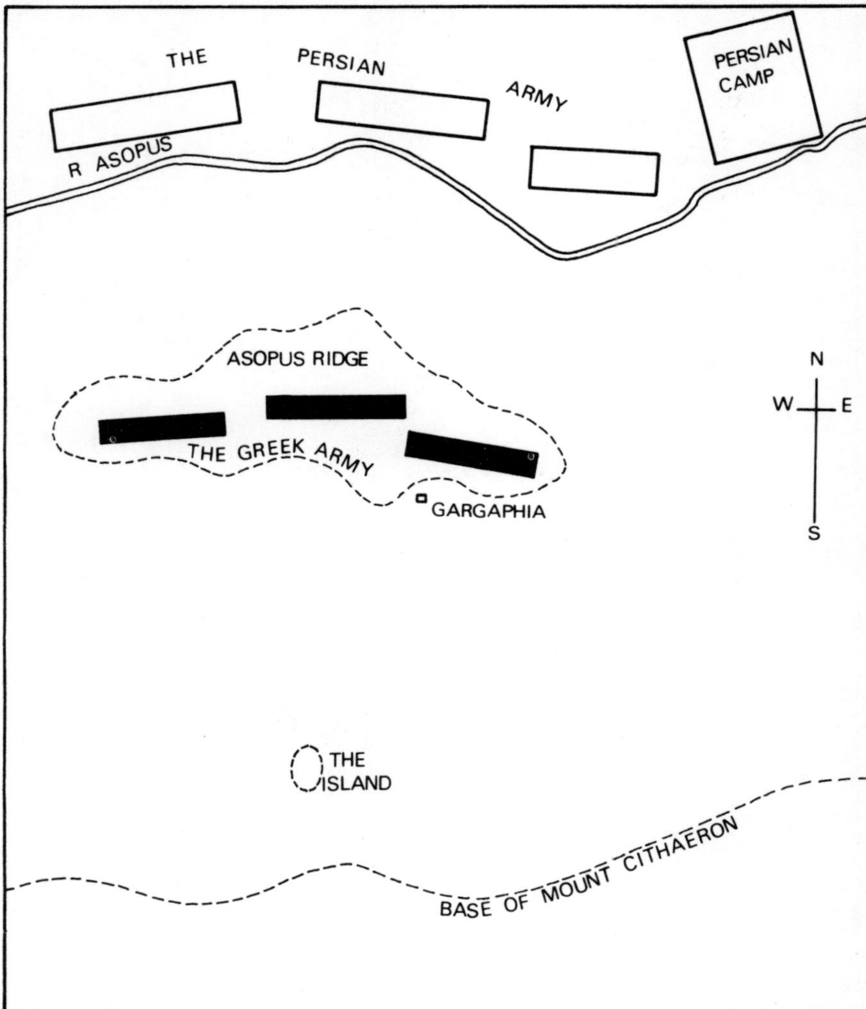

The elements of the Battle of Plataea 479 BC. Map 1: the armies in presence

which shows the outlines of the terrain features relevant to our re-enactment of the battle.

The position of the Greek forces is that which they occupied after leaving the foothills of Mount Cithaeron, where they had already been rather imprudently attacked by the Persian cavalry—a rash undertaking as the ground was rough and most unsuitable for mounted action, and the cavalry in fact received a severe drubbing. Following this success and possibly somewhat elated by it, the Greek army

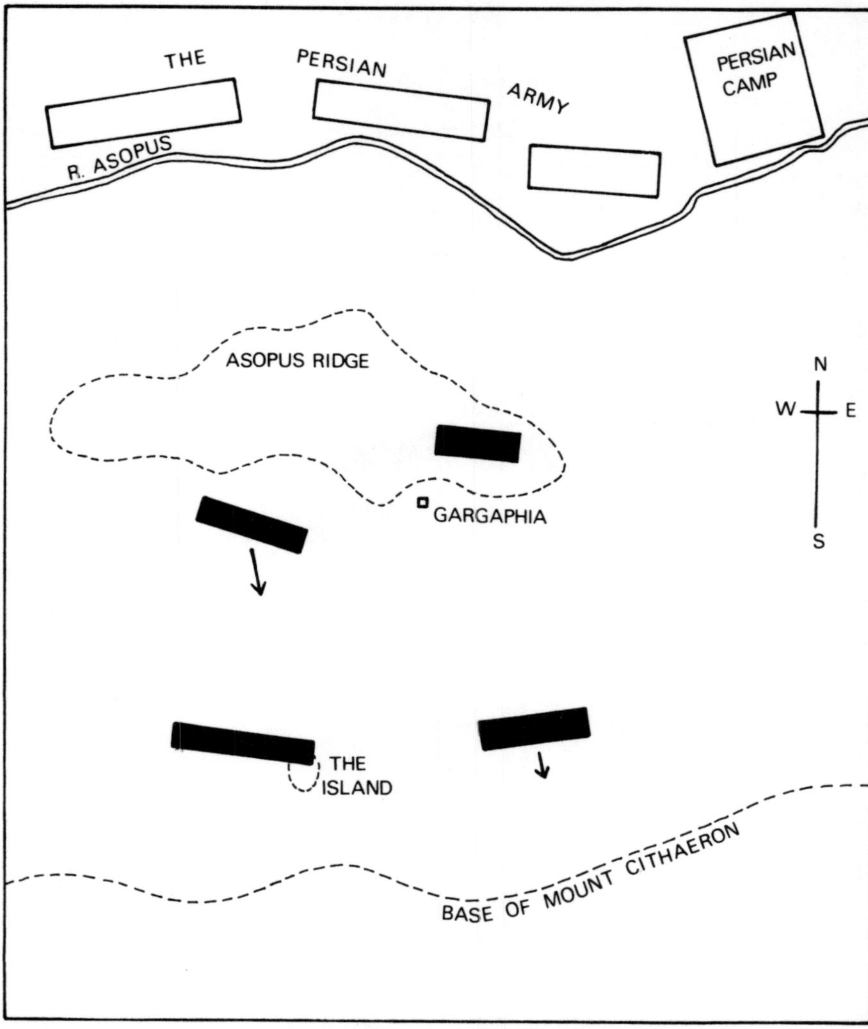

Map 2: probable situation at dawn on day of battle

moved forward to take up the position shown in the map, along a low ridge of hills south of the Asopus river, beyond which were the Persians. The two armies remained thus for eight days, at the end of which, as a result of a successful Persian raid on a Greek provision train coming through the passes of Mount Cithaeron, a council of war was held by Pausanias and his fellow commanders, where it was decided that the Greeks should withdraw to their first position on the foothills of the mountain. This manoeuvre was to be carried out at night and in sections—first one contingent moving off, then another, and so on—but through a sequence of events (some unavoidable and others simply ludicrous) the entire operation was performed in a most disjointed fashion and when day broke, the cohesion of the Greek forces was completely lost, each contingent being isolated from the others. To the Persians, the Greek centre had disappeared and the two wings were retiring over the rolling ground that was so suitable for cavalry. Map 2 shows the ensuing action.

Seeing his enemy thus dislocated the Persian Mardonius without more ado flung his cavalry against the retreating Greek wings, following them up with infantry. The Spartans apparently took the brunt of the Persian attack, the squadrons whirling about the hoplites at a safe distance while plying them with volley after volley of arrows until the infantry came up, their arrows now falling thick and fast upon the Greeks. Hoplite casualties were increasing rapidly but as more and more Persian infantry came up they were packed behind each other in such a mass that their power of manoeuvre was so inhibited that when the Spartans charged the Persian foot was quite unable to retreat. Nevertheless they put up a desperate resistance, fighting most bravely until Mardonius himself was killed, whereupon the heart went out of them and they melted away in rout. Far away on the other flank where the Athenians were posted it was a question of Greek meeting Greek, as they were attacked by the Boeotian and other Greek allies of the Persians. Here again there was bitter fighting until the Athenians prevailed and the 'Medising' Greeks broke and ran. Uncommitted so far, the Greek centre marched briskly forward when it realised what was happening and was promptly attacked by Theban cavalry (in the service of Perisa, of course) and one of its columns was handled most peremptorily, the only really severe losses sustained by the Greeks being incurred here.

Pursuit was maintained as far as the Persian stockade into which the Greeks broke, much slaughter followed before resistance collapsed. The victory was complete and indeed final—the last attempt of Persia to subjugate Greece had been thwarted.

This then is the battle we are about to enact.

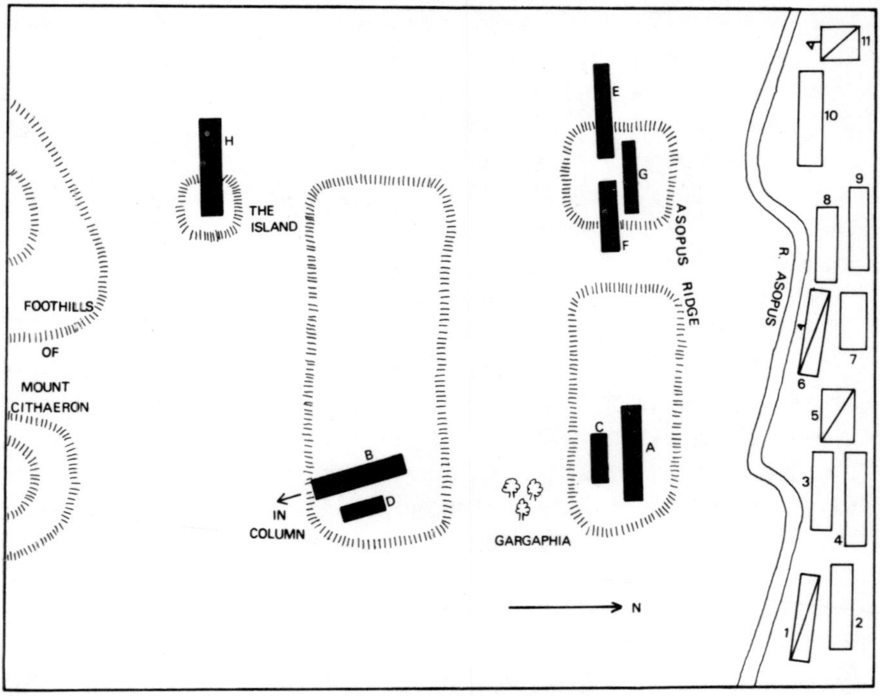

Plataea as a war game. Map 1: the initial situation

Greeks

A—Spartan Alpha Battalion
B—Spartan Beta Battalion
C—Messenian Javelinmen
D—Antelope Javelinmen
E—Athenian Omicron Battalion
F—Athenian Sigma Battalion
G—Athenian Archers

Persians

1—'Horsetail' Light Cavalry
2—'Green' Immortals
3—Median Javelinmen
4—'Black' Immortals
5—'Scorpion' Heavy Cavalry
6—'Crescent' Light Cavalry
7—Cadusii Light Infantry
8—Apadan Light Infantry
9—Phrygian Infantry
10—Boeotian Infantry
11—Theban Light Cavalry

XI

The Battle of Plataea—The War Game

From the preceding chapter it will be evident that Plataea is one of those battles which can readily be sub-divided into phases, each of which can be refought as a separate engagement, thus enabling the student to make a really good study in depth of it. Having done this with other battles in the past, I know that it is possible to achieve a very complete understanding of the relationship of the various tactical situations and of how all the individual sections can be assembled into a coherent whole. For the purpose of the present exercise, however, it seemed that it would be more advisable for several reasons (among which was a lively desire not to inflict undue tedium upon the reader as well as the realisation that to treat Plataea as a series of war games would doubtless be to take up far too much space in the present work) to deal with the battle as a single war game complete in itself, and to stage it at the time of the final day referred to in the previous chapter, when first light allowed the Persians to see the Greeks in the act of carrying out their somewhat uncoordinated withdrawal. To deprive ourselves of a re-enactment of the initial Persian cavalry attack upon the Greek infantry on the slopes of Mount Cithaeron, as well as of the particularly exciting raid upon the waggon train (this always being an extremely good scenario for a war game) was sad, but as events were to prove the decision was a wise one and the war game which followed was an outstanding success from all points of view.

To begin the story, then, we shall describe the ground and, as in a great many ancient battles, there was little in the way of ground irregularities to impede the movement of troops especially as the

Greeks operated in fairly tight formations. In effect, on the actual battlefield there were no steep hills, unfordable rivers, dense woods and so on. Map I in fact shows the debatable ground as it was set up on the war games table which measured nine feet by seven, the size we always use for such major engagements. The Asopus River is of course on the right of the map—the north side, that is—and immediately to its south are the two low hills, divided by a valley, which constitute the Asopus Ridge. On the southern edge of the map are the first indications of the foothills of Mount Cithaeron and between them and the Ridge lay a stretch of gently rolling ground, none of the elevations shown being more than low hills with very slight gradients, not sufficient to afford any obstacle to movement, but of enough height to give impetus advantage to troops charging downhill or fighting against an enemy on a lower level. The small woods are of no great importance, that to the south-east corner of the Ridge indicating the presence of a well—Gargaphia, as it was called. Finally, to the Asopus River itself. Writers tend to be more than a little vague about just what sort of stream this was, how deep or how wide, but certainly at the time of the battle it must have been fairly inconsiderable, as no mention is made of troops having any difficulty in crossing nor is it suggested that any search had to be made for individual fords. So, having weighed up the possibilities, to which was added the fact that as the battle was fought in late summer, water level would in any case have been low, I decided that it was everywhere fordable, although with just a little difficulty, and that the movement of both infantry and cavalry through it would be appreciably slowed down. In any case both would be considerably disorganised during a period spent wholly or partially in the crossing. As we shall see this proved of some importance on one occasion.

Naturally enough, a great deal of consideration had to be given to the number of troops we should use in the game—the actual figures, that is—and of course the strength of each side would have to be correct in relationship to that of the enemy. At the same time enough troops would have to be employed for the players to get the feel of having taken part in a big battle but, at the same time, not too many for the table to be overloaded and for manoeuvre to prove impossible. After some thought, then, it was decided that the Greek army should be composed thus:

Spartan Alpha Battalion (body armour, shield, long spear, sword)	50 heavy hoplites
Spartan Beta Battalion (body armour, shield, long spear, sword)	50 heavy hoplites
Athenian Omicron Battalion (shield, long spear, sword)	42 medium hoplites
Athenian Sigma Battalion (shield, long spear, sword)	42 medium hoplites
Corinthian Battalion (shield, long spear, sword)	42 medium hoplites
Athenian Archers (bow, sword)	20 light infantry
Messenian Javelinmen (shield, javelin)	12 light infantry
Antelope Javelinmen (shield, javelin)	12 light infantry

In actual miniature figures, then, this array amounted to 270, to which we can add three 'general' figures, of whom more later.

At first sight, the number of archers seems rather high, even bearing in mind the importance of realism. On reflection, though, the proportion of archers is not really unduly excessive and indeed events were to show that the number of bowmen could have been halved without having any influence upon the fighting.

The presence of the two small javelin units merits a little explanation. The accounts of the fighting, sparse as they are, concern only the hoplites on the Greek side (and naturally the few archers) and there is no reference to light infantry—peltasts, javelinmen or whatever—in contemporary sources. It is known, however, that a large number of helots—serfs, or indeed slaves—accompanied the Spartans, as many as seven to each warrior according to some authorities, and there is indeed one reference to 35,000 such helots, although this is probably an inflated figure, certainly as far as their fighting value was concerned. Even so, I feel that in this large body of men, who were supposed to assist their masters in every way, there would be at least a small proportion who could, if called upon, act as light infantry in a skirmishing role. To this end then, as the Greek Order of Battle shows, there was allocated to each Spartan battalion a small twelve-man javelin unit. To ensure that, in extremis, their presence would not be crucial, they were given the lowest possible combat morale rating, which meant that their staying power in mêlée was minimal—the simple fact of their receiving in hand-to-hand combat more casualties than they inflicted would cause them to break and run. Their courage was deemed to be of the lowest, next to their being unwilling conscripts in fact.

It is now the turn of the Persian army, which was rather more difficult to estimate in its component details. Finally, however, the following was arrived at, as being the most accurate.

'Green' Immortals (bow, spear, shield)	42 medium infantry
'Black' Immortals (bow, spear, shield)	42 medium infantry
'Horsetail' Light Cavalry (bow, spear, shield)	18 light cavalrymen
'Scorpion' Heavy Cavalry (spear, shield, body armour)	14 heavy cavalrymen
Median Javelinmen (javelins, shield)	20 light infantry
'Crescent' Light Cavalry (bow, spear, shield)	14 light cavalrymen
Cadusii Light Infantry (bow, spear, shield)	21 light infantry
Apadan Light Infantry (bow, spear, shield)	21 light infantry
Phrygian Infantry (spear, shield)	41 medium infantry
Boeotian Hoplites (long spear, shield)	50 medium infantry
Theban Light Cavalry (spear, shield)	20 light cavalrymen
	A grand total of 303 figures.

It is apparent that the totals for the two armies are not strictly accurate as to numerical ratio, but a little juggling was necessary to compensate for the fact that the proportion of formidable medium infantry as supplied by the Greek allies of the Persians—the right wing—was rather high, thus over-increasing the proportion of regular troops in the Persian army. The idea was, as always, to ensure as much realism in accordance with the known facts as was possible, but at the same time to err, if at all, in the direction of not making the war game too one-sided.

Possibly a brief reminder of the weights of the different sorts of infantry may be in order. The heavy designation normally implies the wearing of metal body armour, together with helmet and shield, medium being similar except for the body armour, which would very likely be of leather rather than, say, bronze. Light infantry had no body armour, but might in certain cases have had a light shield, and in the case of javelinmen, possibly a helmet. The Greek archers, incidentally, had no defensive paraphernalia whatsoever. Naturally, the light infantry are more mobile than the others, the heavies, as might be expected, being the most ponderous in movement. Thus, we can see that a considerable part of the Greek army was heavy, while of the Persians only the Scorpion cavalry came into this category.

Obviously, the dearth of source material and the reprehensible habit of ancient historians and annalists of ignoring the most important details—for wargamers especially—of battle tactics, times and movements, make it apparent that only within very general limits can we know what a situation was at some particular time; but as far as we can judge, the position at dawn on the day of the battle was that shown in map 1. The Greek centre has executed its retreat and is established on the little hill, the Island, and part of the right wing, the Spartans, is moving back in column, leaving the remainder to follow along with the left wing, the Athenians. It is not easy to determine just how the various bodies moved in relation to each other, and there may well be differences of opinion on the layout as shown, but it seems to be the nearest approach to what the position actually was. The Persians, of course, were massed along the north bank of the Asopus, ready to cross.

With the troops arranged thus, it might be proper at this moment to specify the generals, of whom there were three on each side. First, the Persians were led by Mardonius, in the person of Derek Casey, who was well supported by Stan and Martin Medrow, the latter commanding the 'Medising' Greeks of the right flank. Overall commander of

The Battle of Plataea: viewed from north of the Asopus, the Persians have crossed the river and are about to assail the eastern end of the Ridge. In the distance Pausanias is approaching

the Greek army was Pausanias, who in fact was Alan Stoneman, the figure representative of him being with the battalion moving back towards Cithaeron. His colleagues were Andrew Green and Andrew Mummery, who commanded the Corinthians and Athenians respectively. Actual identities were predetermined by a lottery, each player drawing a slip of paper bearing the identity to be assumed and the composition of the forces under the particular commander, plus a brief summary of the situation. Certainly at the outset, none was aware of the exact historical battle which was to be re-fought, although certain suspicions were doubtless entertained.

In any event, as soon as dawn broke, that is during the first period of the game, the Persians moved off towards the river, some units, notably the light cavalry, being able to make the crossing at once (they have an ordinary move of 16″, but this of course was considerably reduced for the fording operation). On the Greek left the Athenian archers advanced to the northern edge of the Ridge (which was only a slight elevation) and bent their bows upon their renegade compatriots opposite, the Boeotian hoplites, who incurred thereby a few casualties and consequently had to take a morale test—the rules stating that this must be done when a unit receives its first fire in a game. This was disastrous for the Boeotians. With such an array of support in their vicinity, with apparently only a few archers facing them and even with only an average dice throw, they threw discretion to the winds and went into an all-out charge without orders, splashing through the river towards the bowmen. These fell back nimbly out of reach, and as the Boeotians fell back to regroup they were charged in the rear by the Omicron battalion of Athenians. (This charge without orders lasts two periods, at the end of which the affected unit stands fast to await fresh orders, but in this case an immediate morale test showed that they had to retire—a very reasonable thing, really, in the circumstances). This attack did the Boeotians no good at all and they fled back across the river, having suffered heavy casualties. With orders not to pursue, Omicron halted on the Ridge.

While all this was taking place, very much to the discomfiture of the Persian commander, his forces were crossing the river in strength and on the Greek side Pausanias, with his hoplites and javelinmen as well as the Corinthians, were en route for the Ridge, the Greek generals seemingly having decided to hold this feature rather than continue the retreat already initiated (a wise decision, having regard to the

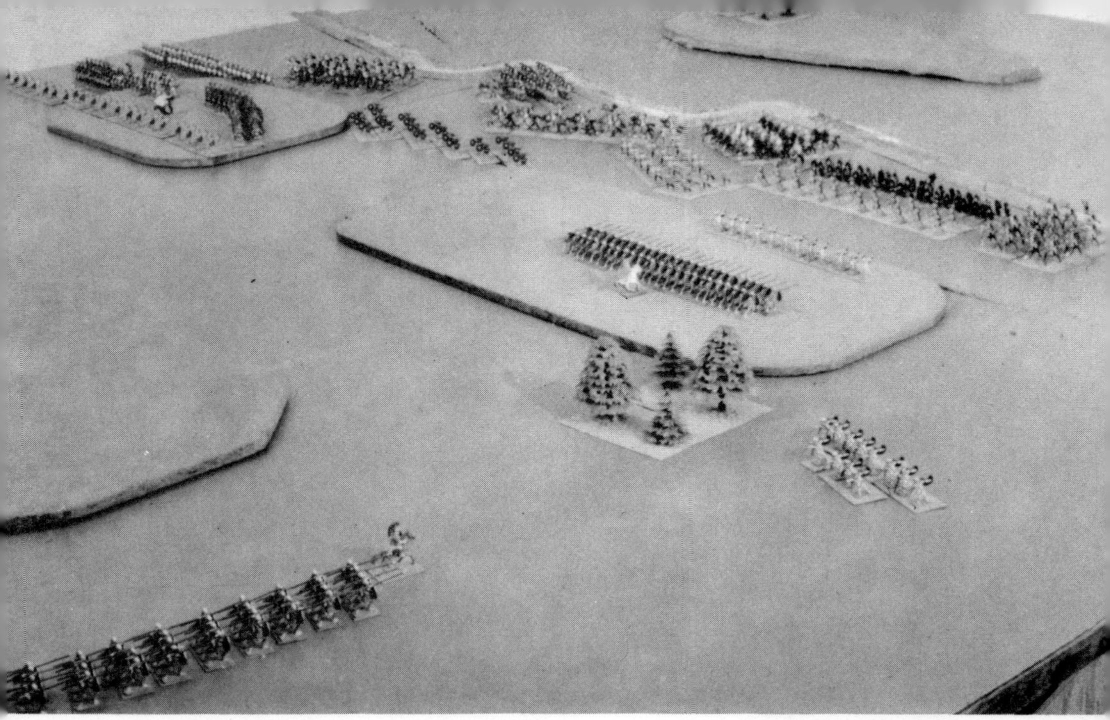

The Battle of Plataea: the early stages as seen from behind the Greek position. Pausanias and his battalion in the foreground

preponderance of Persian cavalry). In the centre the Apadan light infantry, supported by the Crescent light cavalry, were moving towards the Defile with the apparent intention of threatening the right flank of the Athenians. The Messenian javelinmen, covering the front of the Spartan Alpha battalion, were coming under a hail of arrows from the heavy concentration of bowmen opposite, and on successive periods having to test their morale (always done when losses have reached a certain proportion during a period). It was evident, however, that the main Persian attack was going to fall upon the Alpha battalion who, despite their armour, began to lose casualties to the numerous enemy archers, among whom figured prominently the Black Immortals. The Spartans braced themselves for the coming ordeal as their support, the Beta battalion, was still a long way off, having opted to wheel round to come up on the right of the Gargaphia wood rather than turning about to retrace their steps directly.

By the fifth period fighting was fairly general along the entire front and the Crescent light cavalry, ascending the western end of the Ridge, collided with the Athenian Sigma battalion and were flung back in rout (it is most unprofitable for light cavalry to come into contact with the long spears of a formation of hoplites). Fighting on

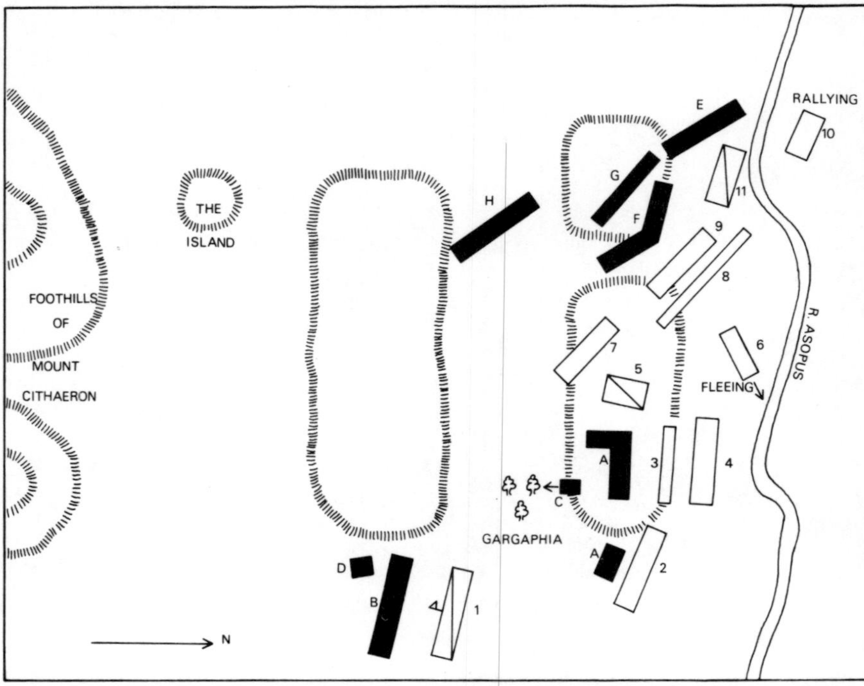

Map 2: the situation at the end of the sixth period

the Persian right flank had so far been all in favour of the Greeks, both the Crescent cavalry and the Boeotian battalion being in flight, but despite what was happening about them, a morale test showed that the Theban light cavalry did not panic but stood fast to cover the hasty withdrawal of their fellows. Still, they too suffered from the fire of the Athenian archers who had come forward again to the northern edge of the Ridge. Gallantly covering the Spartan Alpha battalion, the Messenian javelinmen took heavy casualties from the arrows of the Immortals and when finally attacked hand-to-hand by the much more numerous Median javelins, who had by-passed the Immortals, were first forced back and then under continued pressure what was left of them broke and ran, this being only a few figures.

At the end of the sixth period the situation was as shown in map 2. On the Greek left both Athenian battalions are moving steadily forward to engage the Theban horsemen, while across the river the Boeotian infantrymen have rallied and are in process of reforming; masses of Persian infantry have pushed into the Defile and have swarmed on to the eastern end of the Ridge, where the Alpha battalion is under heavy frontal pressure from the Median javelins while the Black Immortals are firing over the heads of the Medes into the Spartan rear ranks. The battalion has also been obliged to 'refuse' its left flank to face a charge about to be launched by the Scorpion heavy cavalry. The Messenian javelinmen are no longer of any fighting value. A detachment of Alpha is at grips with the Green Immortals just east of the Ridge and the Horsetail light cavalry has ridden past to hold up the advance of Pausanias' men. At this stage, despite the Greek success on their left, the situation was one of extreme peril, the outcome of the struggle being balanced in fact upon the proverbial razor's edge, and the not unexpressive lineaments of the Greek commanders bore distinct evidence of the anxiety to which they were a prey at this stage. The Corinthians were fairly close at hand, however, and would be making their presence felt in a couple of periods or thereabouts.

On the Persian left it took the Horsetails only a single period to drive off the Antelope javelins covering Pausanias' battalion but their own impetus carried them onto the spears of the Greeks and in this collision they were utterly shattered, losing no less than five figures in two periods of fighting and consequently fleeing in disorder. To counter this, though, the small Alpha detachment, after putting up a strong fight against the much stronger Green Immortals finally broke and fled. The main body of the battalion, however, was hanging grimly on to its position on the Ridge, despite combined and simultaneous charges against them by the Black Immortals and the Scorpion heavy cavalry. The Spartans had driven off the Median javelinmen and the cavalry was now forced to retire as well. Even so, Alpha had been pushed almost to the southern side of the Ridge by the Black Immortals who consequently now had the additional advantage of fighting downhill. In the centre the strong Phrygian regiment had come up on the left flank of the Theban cavalry, but charged by the Sigma battalion of Athenians was driven back. During period 7 the Corinthians had also come into the line of battle in the centre, bringing pressure to bear on the Apadan light infantry who for some time had been shooting at the Athenians with impunity.

The battle was now at its climax. The Spartans of the Beta battalion came up on the east of the Ridge to engage the Green Immortals, but themselves were charged in their left flank by the Scorpion cavalry who had, with considerable daring, swept round the rear of Alpha, still engaged with the Black Immortals. Relief was at hand, at last, and

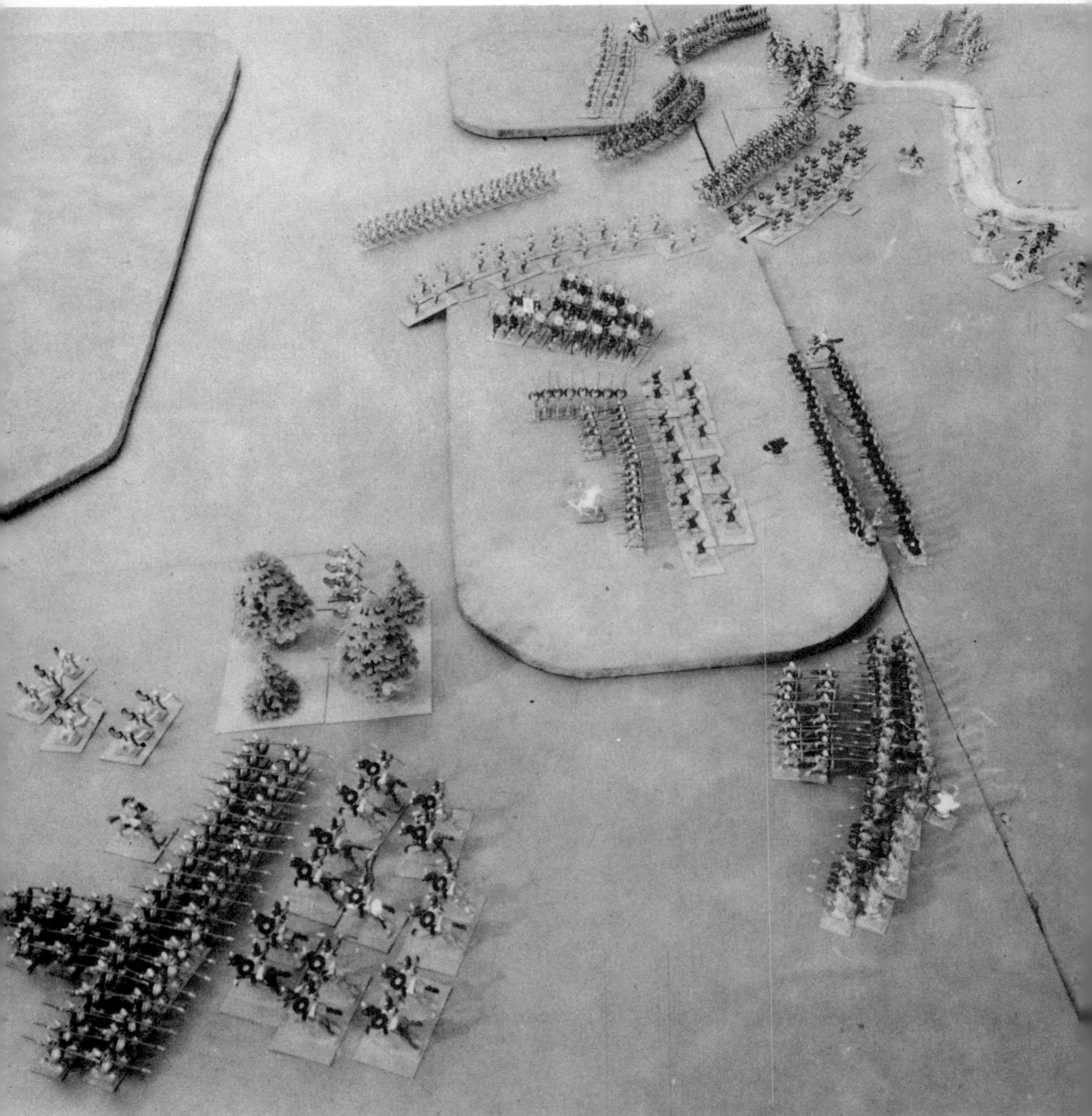

The Battle of Plataea: the climax approaches with the Greeks everywhere under pressure. In the foreground the 'Horsetail' light cavalry are about to be broken by the Spartans and the Greek left wing is beginning to force back the Persian right ascending the Ridge, pushing aside the Apadan infantrymen, who were now caught up in the flight of the Phrygians, the Corinthians charged the right flank of the Black Immortals, who broke and fled before this onslaught. (The casualties they suffered at the hands of both Spartans and Corinthians were appreciably more than twice the number they themselves had inflicted). The Persian centre had now definitely collapsed, many units being in full flight across the river. This was partly due to the spread of panic from the Phrygians. (When a unit breaks and runs, all other bodies of troops within a certain distance must test their morale and obviously the sight of a fleeing regiment does nothing to raise the spirits of any nearby troops). By this time, too, the Theban light cavalry had been pressed into the river by the Athenians, although across the Asopus the Boeotian hoplites had rallied and moved up to the river bank. The Athenian Omicron battalion had been moved down to the south bank, however, to prevent their crossing. (Had the Boeotians attempted this, they would have been so disorganised by the process of crossing that they would have had no chance at all against the waiting Athenians). The situation shown in map 3 is the final stage of the battle. The Black Immortals are about to break, and the Scorpion cavalry has already done so, receiving twice the casualties they had inflicted. (It was virtually unheard of historically for even heavy cavalry to defeat formed infantry with long spears). The Green Immortals had been pushed back—although receiving higher casualties than they inflicted the difference was not sufficient to cause a break, only the push back. With the entire Persian army in flight or at least retreating the Greeks were the undoubted victors and as it was now nightfall (the number of periods fought equalling a full day in time) the Persians made the most of the darkness and kept going without more ado. There was no pursuit, it being considered that the exhausted Greeks lay upon their arms during the night.

It might be interesting to note actual losses, that is the number of figures removed from the table as being battle casualties, as opposed to those still in situ but ineffective militarily because they had been routed. A final count showed the Greek loss to be 67 and the Persians' 74. This is striking testimony to the way good war game rules operate. No longer, as in yesterday's war games, the battle ends with about three figures on one side and none on the other. The result simulates much more realistically the effect of good tactics as well as morale.

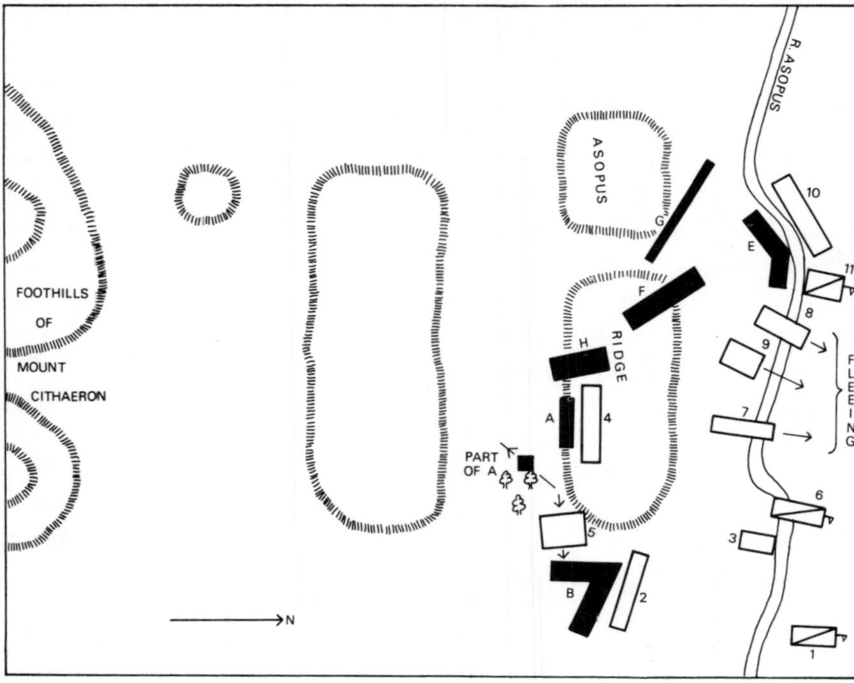

Map 3: the end

So ended the war game version of the Plataea battle, as its prototype had done, in a Greek victory. In addition to this historically correct conclusion it was, in fact, a first rate war game, a veritable cliff-hanger for a great deal of the time and in truth I cannot recall any battle I have fought or witnessed which was more exciting. Still on a personal note it was very pleasing that my interpretation of the terrain and the forces engaged worked out so satisfactorily—all parties involved enjoyed it enormously. What was also gratifying was that the rules operated smoothly and also that neither side had a monopoly of good or bad dice throws, these being very average throughout. It was, in fact, good Greek tactics and the staying power of their troops which won the day, and as history proved on more than one occasion, the lightly armed and unarmoured Persian was no match in hand to hand combat with the heavy metal of the hoplite, especially the Spartan. Though quite fortuitous, the combination of the players on the contending sides was a happy one, each team being composed of a splendid mixture of enthusiasm and experience. Indeed, the ability to seize tactical opportunities was everywhere apparent, and, in truth, it was a great game.

XII
The Macedonians

As has been indicated, Sparta was for long the dominant land power in Greece, but the first half of the fourth century BC saw something of a change, with Thebes establishing a strong claim to this position, thanks largely to the defeat of the Spartans by the Theban commander, the famous Epaminondas, at Leuctra in 371 BC. This was almost entirely due to a tactical innovation for which he was responsible and which was duplicated many, many centuries later by Frederick the Great of Prussia. In simple terms, the Theban general deployed a powerful striking column fifty yards deep on one flank, at the same time refusing the other—in a word, the oblique order. The impact of this mass was too much even for the Lacedaemonians and the resulting defeat contributed very considerably to their subsequent decline in status and power. However, the near-supremacy of Thebes was destined to be of short duration for, in the hour of yet another victory over Sparta at Mantinea in 362 BC, Epaminondas fell, and without his genius the influence of Thebes languished and in fact was rapidly extinguished. There was no Greek state of sufficient prominence to provide any sort of leadership for the country as a whole, but this proved of no consequence for from beyond the Greek border came a force which was eventually to take the entire country and much more in fief.

Lying to the northeast of Greece, Macedonia had for long been looked upon by her southern neighbours as a barbarian nation, although its people were substantially of Greek stock. However, in the twenty or so years after 359 BC, Philip II, the King, by a masterly combination of organisational ability, tactical skill and political acumen, was able to create a most powerful state and army from the highly disparate elements he had in hand. When he moved against Greece—this being the first stage of the plan he had conceived—he totally defeated a combined Theban and Athenian army at Chaeronea (338 BC) and henceforth, in all but name, was hegemon of Greece. The subjugation of Greece to his will was in fact but a mere stepping stone towards his chief aim in life, which was to wage a successful war against Persia. Alas for his dreams, he was assassinated in 336 however, and it was left to his son, Alexander, to carry out his father's ambition, which he did many times over.

Alexander is most justly known as 'the Great' and is much more worthy of the description than many other rulers who have been given or who have assumed such a title. True, he inherited a most excellent army from his father but he improved it out of all recognition, creating such an instrument of war as had not previously been seen. Two swift local campaigns, one against Thebes and the other on the Danube, sufficed to blood his army and without more delay he crossed the Hellespont into Asia Minor on the first stage of a career of conquest which was to take him to Egypt and ultimately to the borders of India. In 334 BC he defeated a powerful Persian army, reinforced by a strong contingent of Greek mercenaries, at the Granicus. For a year thereafter he was engaged in establishing bases, setting up lines of communication and in reducing enemy fortresses, but towards the end of October 333 BC, Darius, King of Persia, faced him with a great host on the banks of the Penarus in what we know as south-eastern Turkey. The Persians were utterly routed, Darius incontinently fleeing the field. During the following two years, which were full of action but which we can treat superficially, Alexander moved through Palestine, occupied Egypt and reorganised its government, then set his face towards Persia. King Darius had occupied himself during this time with raising another army, greater in number than the previous but of very mixed quality. Only the cavalry arm seems to have been of much account and the only good infantry were the Royal Guards and a relatively small body of Greek mercenaries. The remaining foot soldiery were a conglomerate mass of very unwilling levies and peasant conscripts. Darius had also resurrected the corps of chariots, for long inactive, but as events were to prove, these vehicles were ineffective when faced by disciplined troops. The ensuing battle—Arbela or Gaugamela—was another crushing victory for Alexander. Darius fled yet again and was later assassinated, whereupon Alexander found himself, in effect, King of Persia. With his appetite for conquest still unsated however, he took his army even further eastwards, defeating an Indian army complete with elephants on the banks of the Hydaspes (the present day Jhelum). Tactically, this might be considered to have been the greatest of his triumphs, involving as it did the crossing of the river in the very teeth of the enemy. This was the end of the advance, though, and following a near-mutiny and a sit-down strike by his Macedonians (still a powerful element in his army although he had recruited all sorts of irregular troops as well as Persians) Alexander had to retrace his steps. He died at Babylon on 13 June, 323 BC, at the age of 33.

In his handling of troops and in the combining of the functions of the various arms—cavalry, infantry and so on—Alexander was supreme, and his eye for a battlefield was the equal of that of Marlborough and Napoleon. In his major battles he seemed unerringly to determine the weak point of an enemy army, but it was also in the innumerable small campaigns and actions he fought between the larger engagements that he showed to an even greater extent his innate tactical ability, whether his adversaries were fast moving cavalry in the open or determined infantry ensconced upon a rocky crag. Of course, as we have said, Alexander was fortunate in having as a starter the splendid army he took over from his father, but he was not slow to use it to the greatest advantage in all circumstances. It is not always easy to determine to what extent Alexander or Philip were responsible for the innovations which transformed the Macedonian army into the superb fighting force which marched into Asia but whoever was the military architect, it was certainly the best army we have yet seen in antiquity.

So, in considering its composition and quality, let us first start with the infantry, which was organised in the first instance into a standing force of six battalions and which differed significantly from the traditional hoplite array of the Greek city states. First and most important was a tremendous increase in the length of the spear from the nine feet or so of the old type to something in the neighbourhood of fifteen feet. This new weapon was called the 'sarissa' and at the same time, so that it should be used with maximum effect, the depth of the battle order was doubled, from eight to sixteen ranks. Thus, when the sarissas were lowered into the horizontal position, five, possibly six rows of spear points would project beyond the front rank of the phalanx, many more than had been the case with the shorter hoplite spear. The sarissa obviously required both hands for efficient operation so, as a more or less automatic corollary, the old heavy hoplon was discarded in favour of a smaller shield—still round, though—which, when both hands were levelling the sarissa could be slung round on to the back of the owner. Actually this shield—it seems to have been about two feet or a little less in diameter—might well have been of greater psychological than physical value. The advantages of the longer spear, or pike, is obvious; in combat with ordinary hoplites, the sarissa would reach its target before the shorter spear did, although managing a rather unwieldy weapon, even though it was probably very well balanced, must have required a high degree of skill on the part of the phalangites, as the men comprising the phalanx were termed. They had also at this time given up much of the heavy armour, the bronze corslet having been discarded in favour of a thick canvas shirt, although greaves to protect the lower leg were still worn. Headgear was either a small helmet or a 'causia', a wide brimmed floppy hat, and the secondary armament was ordinarily a short sword.

Just as the traditional hoplite phalanx had its weaknesses, so the Macedonian was similarly afflicted, in some cases more severely. The increase in formation depth from eight to sixteen accentuated the difficulty of manoeuvre and drill had to be practised relentlessly—one can imagine the confusion which might have ensued if a formation of men armed with fifteen foot pikes became disordered—it would have been chaotic to say the least.

Consisting of staunch Macedonians in constant training and professional to a very high degree, the phalanx, or the six original battalions—or 'taxeis'—was the hard core of Alexander's army. It was the centre or lynch-pin of his battle tactics but resembled the Greek phalanx in that its rear and flanks were very vulnerable so, as might have been expected, a large number of light troops had to be available to protect these sensitive areas and much use was made of the numerous warlike peoples whose lands bordered on Macedonia and who followed and fought for Alexander with the same devotion and loyalty as did the native Macedonians. From these Thracians, Illyrians and others came a splendid corps of light infantry, some of whose names occur repeatedly in the accounts of the campaigns. The Agrianian javelinmen are a notable example, and he also had both Macedonian and other archers, some recruited from far-off Crete, these latter

The army of Alexander the Great. A battalion of the redoubtable phalanx and cavalry moving past in support

being renowned throughout the ancient world for their expertise with bow and arrow. From Greece itself came other light infantry, most being the peltasts we have noted already, capable effectively of fighting at close quarters as well as in the rôle of skirmishers. All these served to screen and protect the phalanx, but there was yet another kind of infantry which has still to be described with complete certainty, many authorities differing sharply in their views of their equipment and function. These were the 'hypaspists', who were occasionally referred to as the 'silver shields'. One can only speculate upon them and attempt a very diffident assessment as to what they actually were. It is unlikely, for instance, that they were armed with the sarissa and because they are, from time to time, described as moving at a fast rate, could not have been very heavily armoured. They were frequently called upon for special mission and always appeared in the line of battle where their function was apparently to act as a sort of link between the phalanx and the other arms, notably the cavalry. There seems to be a good argument for believing them to be a kind of hoplite without heavy armour and their principal quality is that they were an elite body called upon when a special effort was required.

Over and above the phalanx and the light infantry, a large body of Greek mercenary hoplites formed an important part of the Macedonian army, their arms and equipment being as we have already described.

Thus far what we have spoken of derives in essence almost directly from the old Greek hoplite army although with the significant changes discussed, but it is in the cavalry arm we find the new element in Greek warfare. Alexander excelled in the use of cavalry, using it as his principal striking force. His mounted arm stemmed from the old Royal Companions formed by King Philip and numbered some 2,000 men. They differed from their Greek predecessors in that they were heavy cavalry, being well equipped with bronze corslet and helmet, and being armed with stabbing spear and sword. At this stage they did not carry shields and it must be pointed out that the horseman still did not have the benefit of stirrups, his spear being used for the free-hand stab as he could not use it in the couched position. Nevertheless, the charge of such a body of horse, knee to knee in a disciplined formation was a most important factor in Macedonian tactics and indeed was in the main the means by which Alexander won his victories. This Macedonian cavalry can be considered to be very much an elite body and had many privileges, although in addition to the Companions Alexander had a strong body of non-Macedonian horsemen, mainly recruited in Thessaly and who were virtually as good as the Companions themselves.

Light cavalry, of course, was available in large numbers, Thrace providing much of it. As the Asiatic invasion progressed, however, large numbers of local cavalry were recruited, mainly Scythian in origin, and being armed with the composite bow provided an ideal skirmishing element. Also, but in smaller numbers and possibly only on an experimental basis, Alexander provided a section of his light cavalry with the sarissa, maybe a slightly shorter version of the infantry weapon, calling them 'sarissaphoroi'. It must be remembered that the sarissa was a two-handed weapon and for this to be used on horseback by even the most skillful and experienced rider would require a very considerable effort. The introduction of shields for cavalry seemed to have caused the disbandment of the sarissaphoroi after Alexander's demise and the experiment, if it were so, does not appear to have been an unqualified success. Incidentally, wherever possible the commanding officers of these allied units were Macedonians.

As we are concerned in the main with field operations—especially from the war games point of view—we need say only that Alexander's siege methods measured up to his overall military skill and that he undertook virtually every sort of operation from the formal siege of Tyre lasting some seven months and which was stormed in August 332 BC to the spectacular *coup de main* with which he stormed the Sogdian Rock in 327 BC. Space also precludes any lengthy consideration of Alexander's ancillary services which were of a very high order and there is no doubt that, although of an informal constitution, his staff operated most smoothly and efficiently.

Every one of Alexander's great battles shows his complete mastery of tactics although for interest his small actions and lesser known campaigns provide some of the most exciting moments. In these smaller engagements as well as in the great encounters one sees all his forces used to maximum effect. His battle plan usually involved an advance by the phalanx, although there was no intention that it should do other than 'pin' the enemy and when the latter was disorganised and probably in some confusion he would unleash the heavy cavalry—customarily with himself in the lead, and drive through the opposing line where his sure eye had discerned a weakness. The exception to this rule was his last battle, the crossing of the Hydaspes river in 326 BC, when he found the further side of the great stream defended by the Indian King, Porus, with a large army. Once his army was fully engaged, Alexander found that his cavalry had been neutralised by the enemy elephants—a large force of two hundred or more—and it was left to the infantry to defeat the Indian foot after some heavy fighting. Tactically, though, the crossing of this wide river in the face of a powerful and determined force was one of his greatest triumphs.

It is therefore apparent that, with the various types of troops available, a Macedonian army such as was commanded by Alexander provides a most doughty weapon when transferred to the war games table. Everything is there—heavy and light cavalry, infantry of all types, missile men from slingers to javelinmen and archers. From a practical wargaming point of view, though, one point should be noted. It would not be really possible on a small scale to operate a Macedonian army in the correct proportions of phalanx, heavy cavalry and so on—it would be neither realistic nor effective. In a small scale game a tiny phalanx, a handful of heavy cavalry etc. simply would not work. It is a common mistake to scale down a successful army to a point where the proportions would be unreal in a small force. Alexander certainly did not organise special task groups in this fashion—he would, for example, take his hypaspists as a base, add a few hundred cavalry, and possibly accompany this with a body of Agrianian javelinmen—it all depended upon the objective he had in mind. So, if one wishes to reproduce an Alexandrian army of all arms at its peak, then the war game must be a fairly extensive one with armies certainly of 250 to 300 figures as a minimum—or such is my own feeling on this point.

Incidentally I assure the reader that the prestigious elephant has not been summarily dismissed from these pages by the brief reference above to their use by the Indians at the Hydaspes battle—much more will be said of them at the proper time.

XIII
The Successors

The fragmentation of the Alexandrian empire which followed its founder's death was inevitable in view of his having left no viable heir, and there ensued a long period of almost continual warfare. The generals who had been close to him or who had been created viceroys of various provinces fought bitterly among themselves, either to achieve sole rule over all the vast territories conquered by the Macedonian armies, or more modestly, to maintain their hold on the districts they already controlled. These 'Successors' to Alexander, the 'Diadochi', were indeed a spectacular group of banditti—on the grandest possible scale, admittedly—but this is basically what they proved themselves to be, although they all finally proclaimed themselves kings or were accepted as such. Lysimachus ruled in Thrace, Antigonus 'One-eye' had Anatolia and Syria and Cassander seized Macedon itself, but it was Ptolemy, who most astutely assured himself of the mantle of the Pharaoh of Egypt, and Seleucus, who took over much of what had been the former Persian Empire, who were able to establish dynasties of any degree of permanency. Warfare and diplomacy—or the double-dealing which was the customary substitute for the latter—were carried on unceasingly. At the battle of Ipsus in 301 BC Antigonus, then over eighty years of age, faced the combined armies of his rivals and was defeated, the veteran, fighting to the end, finally falling with his face to the enemy.

The battle of Ipsus is interesting from one particular point of view and harks back to the mention of elephants at the Battle of the Hydaspes when, it will be recalled, Alexander found his cavalry of no avail against Porus' elephants. This neutralising of an enemy mounted arm was in fact the special function of these huge animals and we should at once dismiss the notion that they went careering into action like living battering rams, to force their way through by dint of weight and impetus. That they acted as armoured fighting vehicles in ancient warfare seems to have been far from the case and although many generals favoured them and tried to employ them in an aggressive role, their successes were not all that numerous. Ispus, indeed, gives us the perfect example of how they should have been used. In the battle, Antigonus massed his cavalry on one wing under his son Demetrius, who charged and without much difficulty defeated the horsemen of Seleucus, driving them from the field. However, as has happened many times since, the cavalry, either by accident or design,

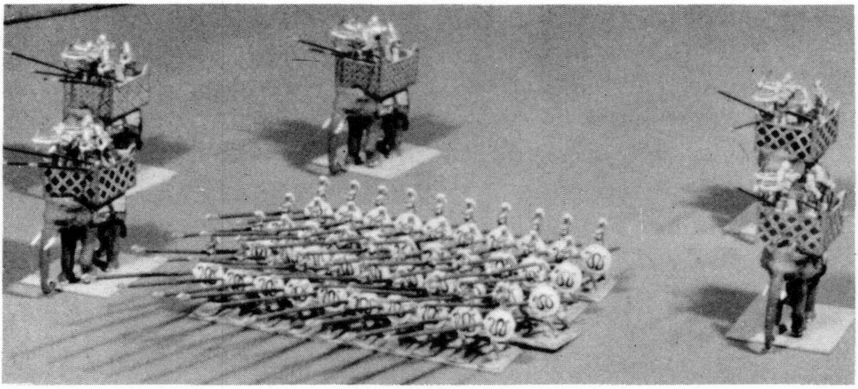

The Successors. The heavier type of phalanx operating in conjunction with war elephants

pursued too far, and when Demetrius rallied his troops and led them back to the relief of his father, whose phalanx was now at grips with that of the allies, he found his way barred by a long line of elephants. The number of these animals deployed in this way varies considerably according to the sources consulted, but they certainly numbered a couple of hundred and may have been substantially more. Demetrius' horses absolutely refused to approach this elephantine wall and no amount of persuasion could prevail upon them to get close to the terrifying odour of the elephants. When the cavalry attempted to circumnavigate the end of the line it simply shuffled further along to counter this and Demetrius could make no headway whatsoever. As we have said, Antigonus was killed, and Demetrius was reduced to waging war at sea. After the death of Cassander he did take possession of Macedon but in a short time lost what was a rather precarious hold upon the country and ended his life as a prisoner of Seleucus.

Life, in fact, was by no means certain for any of these would-be monarchs at this time. At Korupedion in 281 BC, Lysimachus was slain in a battle against his former ally, Seleucus, although Lysimachus was fortunate, in a way, in resembling Antigonus in living long enough to become an octogenarian. In the following year Seleucus himself fell victim to an assassin and as a result of this and similar upheavals there were, by the beginning of the third century BC, only two really dominant powers in the Middle East, these being the Seleucid Kingdom and the Egypt of the Ptolemies, in addition to which rose and fell a number of smaller states of no great consequence, whose borders changed with bewildering frequency over the years.

Warfare as waged by the Seleucid kingdoms is particularly interesting, despite the fact that they made no great contribution to the military art, but nevertheless their colourful armies do merit consideration. Two main qualities characterised these Hellenistic armies, the first being their use of elephants, and it might be a good idea to devote a little time to a consideration of how these animals were employed in action.

First of all, one or two misconceptions are common, one of the most prevalent being the picture of an enormous creature with a vast affair like a medieval keep on its back, this being crammed with numerous soldiers, while the animal itself is covered with armour and is provided with a variety of spikes projecting from every possible point of its anatomy. The truth is far from this, for if we pursue the obvious course and start from the very beginning—not altogether a bad thing—elephants as first encountered by the Macedonians do not appear to have had any castle or howdah whatever. The crewmen, generally the driver and a couple of javelinmen, sat astride their enormous mount as they would have done had they been ordinary cavalrymen. This could not have been the most comfortable of positions, especially as the Indian elephant—to the discomfort of the crew—had a more convex and bony back than the African one. It may be of some relevance to point out that, at this time, when we speak of the African elephant, as we shall do, we do not have in mind the huge species of today but a smaller variety formerly indigenous to North Africa and which I believe is still found in the Congo, where it seems to be domesticated by the natives. The Indian elephants used by Porus would have had their tusks, if any, reinforced with points and rings of metal.

Now, although Alexander's infantry had been able to deal most effectively with Porus' elephants at the Hydaspes battle, the beasts undoubtedly made a great impression on the Macedonian commanders, notably Seleucus, who later exchanged a large tract of his newly acquired territory for a supply of Indian elephants. Now, in true Macedonian fashion Seleucus and others of his ilk set about improving the fighting potential of his elephants and quickly added a fighting tower to the standard equipment (this article, it seems, had formerly been used by the Indians but only by commanders and kings, presumably as a sort of command post). The towers accommodated as many as four fighting men—this in addition to the driver or mahout, who continued to sit astride the animal's neck.

The armament of such crews undoubtedly varied but obviously, perched high on the elephant's back their weapons would have to be either missiles or very long spears. The latter may well have been used but I think it probable that javelins would have been carried more often and that archers would also form an essential part of an elephant crew. As the Seleucid era (Seleucus had the good fortune to have had his name adopted by posterity as descriptive of the period in which he and his heirs were active) wore on, it might also have been that some ornamental housing would have been devised for the elephants, possibly some stout quilted covering—I do not think there would have been much metal protection, the amount required and the expense involved would militate against this.

Now, the tactical use of elephants varied a great deal. We have seen one classic instance of their employment against cavalry at Ipsus and Indian generals apparently used them in this static sort of role, much as they did at the Hydaspes, but many of the later exponents of the elephant may have tried to employ them as a kind of 'break-through' weapon and as such they had but indifferent success. Certainly against properly trained and disciplined infantry they did not seem to 'come off'. As time went on this indirectly might have been one of the reasons for their use, as really well trained and professional infantry were becoming more and more difficult to come by, plus the fact that the terrain of the average eastern battle favoured the use of elephants and consequently the nations of those parts tended to use them in numbers. The Persians were something of an exception to this rule, however, despite their living in such close proximity to the suppliers. Persian armies had a high proportion of cavalry and the well known effect of elephants on horses would be noticeable. Even if the latter were accustomed to the presence of the larger animal, they would still be more than a little nervous when elephants were at hand. Another characteristic of the Achaemenid Persian military machine was the large numbers of light infantry which made up the bulk of their armies, with which it was not easy for elephants to keep up. We shall be mentioning elephants again as we progress, but at the present time it might be of some advantage to examine the manner in which they should be treated in the war game. It is not the easiest thing to establish rules which give them their proper and realistic place in the line of battle, but a few tentative proposals may be made.

Again let us forget the suggestion that the elephant be treated as a kind of tank, charging with great speed and relative invulnerability. They are, and were, highly nervous beasts and their drivers had to be very much on the alert to prevent their getting out of control, which could happen fairly readily. It seems that rather than crashing through enemy infantry their function was simply to disorganise such troops. It would indeed be infantry which would ordinarily face elephants as no intelligent general would dream of allowing cavalry to endure such an ordeal. In one sense, of course, it is true that elephants could be used as modern armour in that if infantry ranks were opened to allow advancing elephants to pass through the gaps—this being a standard tactic—supporting infantry could follow up their elephants, rush in and exploit the gaps in question.

It was naturally the aim of the defending force to discharge as many missiles as possible at the elephants as they approached, but as far as archery was concerned, it does not require much thought to conclude that this was far more advantageously directed at the crew and driver, there being no profit in shooting at the elephant itself, arrows doing little more than irritating the pachyderm. Javelins seem to have been much more effective, one hitting a tender spot—round the mouth or eye—not doing the animal much good at all.

It would seem essential then to have a rule governing the elephant's behaviour when it was the target of such missiles and I feel that to kill one with, say, arrow fire, the number of hits would have to be very high indeed, although javelins would probably count rather more

Dealing with elephants; archers fire upon them from convenient high ground and javelinmen press boldly up to hurl their missiles

heavily than arrows and the like. As a matter of fact, in default of the actual demise of the elephant, it would happen more often that the cummulative effect of the hits would have another result, that of causing the animal sufficient pain and annoyance to make it break its driver's control and stampede. Once this panic total has been reached, obviously some sort of morale assessment would have to be reached, and if this showed the animal to be likely to stampede, a decision would have to be made as to the direction in which it was going to go. This would naturally be a complete matter of chance and could be decided by a single dice throw. I do not forward this as an inflexible rule but merely suggest that a dice throw of 1 would indicate that he goes forward, 2 and 3 that he goes to the right, 4 and 5 to the left, and 6 that he turns sharply about and rushes back in the direction whence he came. All this could be included in the body of morale rules and, of course, if the panic-stricken animal was in the middle of a line of his fellows, into one of which he cannons in his flight, then that one will also have to be diced for to determine whether or not he was affected by the collision. This could be done by a combination of dice throw as well as, for example, a consideration of the number of missile hits he had suffered, and the effect upon him of the crash of the other elephant against his flank.

Apart from orthodox tactics, it should be possible for certain types of army to provide other defences, such as caltrops—arrangements of four iron spikes which, when the device was thrown down, was so arranged that one spike always projected vertically upwards. In some circumstances ditches could be dug, and left open or, if the material was available, could be covered with brushwood or something similar to conceal their presence and trap the ponderous and possibly unsuspecting beasts. Large stakes could also be set into the ground pointing towards the approaching elephants.

Although far from being exhaustive the preceding should give some indication of how the elephant acted in battle and how, in theory, he could be dealt with. In war game terms the elephant can be dangerous but is by no means omnipotent, as I have seen demonstrated in numerous games. Actually, when elephants are fielded in battle, the result can be very spectacular—the most I have personally observed was a vast force of no less than 25, vast in war game terms that is, but it was in fact shot to pieces by massed archery and fled in confusion, dragging the remainder of the army along as well.

Leaving the elephants, now, we have to note the other characteristic of a Seleucid army—the phalanx, which increased in size and deteriorated in quality as the years passed, despite experiments such as the increase in length of the sarissa to an incredible length, up to 24 feet it is said. The proportion of phalangites to the entire army rose dramatically, really to the detriment of the heavy cavalry component, which was rather neglected. As the size of the phalanx increased, so it became more and more unmanageable although, it is recorded, its appearance to an enemy was still quite terrifying. Even so, it was no longer the force it had been under the great Alexander, and when it was met by as highly trained, professional and flexible a force as that fielded by Rome, its inferiority became startlingly obvious. Nevertheless it is just to say that it had made its mark upon the ancient art of warfare, equally as did the Roman legion which was to succeed it.

XIV
Rome and Carthage

Emphasis must now shift substantially, westwards and away from the Eastern Mediterranean where we have lingered so long, and we must simultaneously retrace our steps in time—as there has been some overlapping—briefly to study the beginnings of another massive confrontation, similar in many ways to that which had already occurred between Greece and Persia. As in that case, so in this new one, one side triumphed and the other collapsed, although there was a significant difference in that here the victor flourished, politically at least, much longer and more influentially than did Greece. The confrontation with which we now concern ourselves involved the imperiums of Rome and Carthage. It is without profit to delve into the origins of Rome other than to say that the date of its legendary foundation is 753 BC and, as the centuries passed, what had been originally but one of a number of small Italian city states, gradually, through the economic process and the innate dynamism of its people, waxed mightily in power and extent, despite several defeats by northern barbarians, including the actual burning of Rome itself in 390 BC. Thus, by the beginning of the third century BC Rome had mastered almost all of Italy and the scene had been set for a most bitter struggle which was to last for well over a century.

The antagonist which was to cause the Roman state to put forth every ounce of energy was Carthage, whose origins can be traced back to a period even before those of Rome, in fact to the foundation of trading settlements by Phoenicians from Tyre, who established themselves in what is now Tunisia, their capital, Carthage, being actually founded in the ninth century BC. This city, with the great advantages

The Roman Legion. Three cohorts of this formidable infantry with legate and eagle

provided by the trading skill and commercial acumen of its people, rapidly grew in wealth and status until it was dragged, possibly not too unwillingly, into war with Rome. The first Punic War (265 to 241 BC) was in fact triggered off by both Rome and Carthage's dabbling in the tangled web of Sicilian politics. A number of one-time Greek colonies were located in the island and around the southern tip of the Italian peninsula and one of their contending factions ill-advisedly appealed impartially for aid to both Rome and Carthage, thus drawing both these expanding states into the lists.

Even by this time Rome, although her military system was still very much in its infancy, was showing signs of the power which was ultimately to make the 'pax Romana' the normal writ throughout the greater part of the ancient world. Her organisation for war was in effect not unlike the levy of the old Greek cities, the army being composed of citizens whose right and privilege it was to bear arms, and their unit, when they were assembled, was the legion. Considerable progress had been made from the old concept of the phalanx, to which the early Romans, it seems, had adhered. What was now operated was, in fact, a sort of tripartite organisation, the legion fighting in three distinct lines of infantry, the first being the 'hastati', made up the youngest men of the levy armed with throwing spears, then came the 'princips', armed with long thrusting spears, as were the men of the third line, the 'triarii', who were the veterans, as far as age was concerned at least. One presumes that all had as much armour as they could afford, the richest naturally being the best accoutred, partly with ring mail shirts, although basically all were equipped with a large oval shield made of wood with a central metal boss, and partially reinforced with metal. The throwing spear of the hastati is worth attention. It was the 'pilum', which was to become the characteristic weapon of the Roman soldier for many years. In effect it was an extremely weighty javelin, comprised of a thick wooden shaft with an iron blade sunk deeply into its length, making it a highly deadly implement in trained hands. It had, due to its weight, a maximum throwing range of not more than twenty yards, possibly less, but at this close range it could penetrate wooden shields and most armour also. In fact, if it did lodge in an enemy shield it was practically irremovable and naturally its weight dragged the shield down, making it impossible for its user to wield it effectively. The result was that the shield had to be discarded, leaving its owner open to immediate attack. The Roman technique was, in fact, to move up to close range, hurl the pilum, then to come to close quarters with the short stabbing sword which was the secondary armament of the legionary, thus taking advantage of an enemy still reeling from the impact of the volley of heavy javelins.

The three legionary lines were themselves subdivided, each into ten companies or 'maniples', and in battle formation they were drawn up in chequerboard fashion, so that the intervals between the maniples of the front line would be covered by the maniples of the second, while the intervals of the second line would be similarly covered by the third. The three lines, incidentally, were several ranks deep and in addition to the basic heavy infantry, each maniple had its attached light troops—'velites'—armed with a number of light javelins. These were, of course, not the heavy pila of the hastati and had consequently a much greater throwing range. They were of the greatest possible use in skirmishing with an enemy prior to the legion's closing with it. Certain light cavalry were also attached to the legion and, all in all, the organisation we describe was plainly greatly superior to anything we have as yet encountered, having all the weight of the phalanx with nothing of its rigidity. The manipular plan made for an extremely flexible formation and the mutually supporting qualities are evident. Obviously there had to be a great deal of training and practice to operate this system successfully, but for many years it provided the backbone of Roman tactics. Over and above the attached velites, the legion was adequately supported by missile men, mainly mercenary slingers and archers. In contrast to the phalanx which, when once committed to fight in close combat, could not possibly break off the fighting without imminent risk of disintegrating, the legion could, on the other hand, relieve its sections with comparative ease. It seems that the Roman theory envisaged that 15 minutes was enough for men to fight before relief was provided by the second or third line as the case might be.

To this highly organised and entirely national system, Carthage provided an almost complete antithesis, for throughout their history native Carthaginians were in a considerable minority in their armies, which depended largely on mercenaries as well as on allies, all of whom were recruited from such territories as were occupied by Carthage or were under her influence. These, at times, varied from Numidian light cavalry to Gallic tribesmen and, in contrast to the armies of Rome, there was a high proportion of cavalry, both light and heavy, in Carthaginian forces. In spite of the disadvantages inherent in this system the native guile of the Carthaginian commanders frequently outweighed the tactical benefits the Roman army enjoyed, for indeed the former were renowned for devious behaviour, much to the discomfiture of the slower thinking Romans, in the early days certainly. The first Carthago-Roman conflict was one of mixed success for both sides, much fighting being carried out at sea—an unfamiliar element for the Romans. The latter were not dismayed, however, but built a navy which behaved so well that it was finally possible to ship an army to Africa, its commander being one Regulus. Very unfortunately for him, as it happened, to the aid of the Carthaginians came Xanthippus, a Spartan, with a force of Greek mercenaries. He was a very skilled general and was able to force his own choice of battlefield upon the invaders, defeating and capturing the Roman general. For the time being the Romans abandoned their African adventure and concentrated on building up their fleet. After a battle in 242 BC in which the Roman navy was victorious the Carthaginians made peace, abandoning Sicily the following year.

The battle wherein Regulus was defeated provides a good example of just how a Carthaginian army of this period could be composed and how great could be the diversity of its component troops. The Greek contingent were doubtless almost entirely hoplites and would fight as a phalanx, if not actually cheek by jowl certainly alongside native African troops of several kinds, while also represented would be Spanish infantry and cavalry. As time went on, of course, we know that Gallic tribesmen would add their élan and ferocity to the Carthaginian hosts. Such an army is popular with wargamers, particularly as it includes such a great variety of types, some being most efficacious on the war games table. I have seen Moorish light horsemen, armed with javelins, literally run rings round much heavier and apparently better armed formations, reducing them to complete or near impotence. To all those listed above, the traditional elephant can also be added, the resulting array resembling in type and tactics certain of the armies which fought under and against the Seleucid monarchs. However, as was proven many times in the Middle East, such an army required a general of more than ordinary merit to ensure that it acted coherently and produced its maximum effort at the proper time. It was the Second Punic War which was to produce such a leader, one who, before he could be effectively countered, probably gave more trouble and anxiety to Rome than did any other of the notable generals who matched their forces against hers, and they were by no means few in number.

During the years following the First Punic War Carthage made great efforts to consolidate her power in Spain, where the River Ebro had been designated as the dividing line between the spheres of influence of the two powers by a treaty to which they had both seemingly willingly subscribed. As might have been expected, provocation and counter-provocation frequently took place and in 218 BC a young Carthaginian general of a noble house led an army northwards from Spain towards Gaul and ultimately Italy. This was Hannibal—truly a name to conjure with in military history. His army, it must be said, was the normal polyglot Carthaginian one, about 40,000 strong—although this figure might well be incorrect by several thousand either way. He had some 40 elephants, and his forces were increased as he advanced by Spaniards, and later by Gauls, and it would seem likely that much of the enthusiasm of these latter two lay in their hopes of plunder and personal gain in the enemy territories through which they would pass. Nevertheless, Hannibal did have good troops with him in considerable numbers, not least the splendid Numidian mounted javelinmen, although these and other types indigenous to the southern shores of the Mediterranean must have suffered considerably when Hannibal's army forced its way through the Alps and finally into the north Italian plain. Three great victories over Roman armies were to follow in the next few years—the Trebia, Lake Trasimene and Cannae. In every case victory was due to the tactical genius of Hannibal who used every device involving surprise and every arm of his army in an astoundingly effective manner.

Alas for Hannibal, however, both time and the enormous resources of Rome were against him and finally the ideal type of general to counter his quicksilver genius took the field against him. This was Quintus Fabius Maximus, known as 'Cunctator' (The Delayer), who simply refused to join battle, waging instead a long drawn out war of attrition against the Carthaginian. This situation lasted for several years, this being greatly to the advantage of Rome, for while containing Hannibal in southern Italy, she transferred the focal point of her strategy to Spain. Between 210 and 205 BC, Publius Cornelius Scipio, later to be known as Scipio Africanus for reasons which will be made plain, completely cleared the peninsula of Carthaginian troops, although not without a great deal of very hard fighting. Once this subjugation had been achieved, it became the turn of Africa, where Scipio landed with a strong army in 204 BC. In the spring of the following year he heavily defeated the Carthaginians on the Bugradas plain, using his cavalry with a skill so far quite unknown among Roman generals. Hannibal was now recalled to Africa and in 202 BC was fought the battle of Zama, the last great engagement of the Second Punic War. Each army probably numbered something like 40,000, but overall the Romans had the greater experience and, for once, superiority in cavalry. To counter this, Hannibal had about 80 elephants, and at the beginning of the battle he attempted to use the beasts in an aggressive role, sending them directly forward against the Romans. As they advanced, however, they were greeted by a tremendous blast of trumpets and a great concerted shouting. Many panicked, others ran into volleys of javelins and to avoid them veered into their own cavalry, creating chaos wherever they went. The Roman cavalry thereupon charged, driving both wings of Carthaginian horse from the field, and the fighting now became really grim, infantry falling in thousands on each side. Finally Scipio drew back his disorganised infantry, reorganised both princeps and triarii into a single formation, and charged yet again, engaging Hannibal's élite infantry, the veterans who had come with him from Italy. For long the result hung in the balance but at the crucial moment the Roman cavalry returned from their pursuit, charging directly into the rear of the Carthaginian infantry. They were utterly destroyed, although Hannibal himself succeeded in escaping.

Zama is a fascinating battle, both from a purely historical and from a wargaming point of view, and, of course, it is with the latter we are most concerned. It is most interesting to speculate how a present day 'Hannibal' or 'Scipio' might decide to fight it. Would the latter open his ranks as did the Romans to meet the advance of the elephants? The initial Roman deployment at Zama was not, indeed, the customary chequerboard formation, but one where the maniples were drawn up one behind the other to allow the elephants which did press home their charge to be channelled along the lanes between the maniples. Suppose, however, the war game 'Hannibal' held his elephants in reserve to incommode the more numerous (two to one, in fact) Roman cavalry? Or if his cavalry had been driven off would he have deployed the elephants to keep the cavalry away as did Seleucus as Ipsus a hundred or so years earlier? There is only one way to decide such questions, and that is by fighting the battle on the war games table. Allowance would have to be made for the superior battle training of the Romans vis-à-vis the bulk of the Carthaginian army. This could readily be done by increasing Roman strength somewhat or, better still, by giving a proportion of the number greater fighting power than that enjoyed by the Carthaginians. Little in the way of terrain would be necessary for a re-enactment of Zama, the battlefield, although its exact location has never been established, being apparently a flat plain.

The empire of Carthage ended, in effect, at Zama. The people were disarmed and it was not long before the Third Punic War, a comparatively brief affair, ended with the actual destruction of the city of Carthage itself. There is not a great deal to be admired about the Carthaginians and many of their mores are to be deplored, but they certainly caused Rome to strain every nerve to ensure a victory. It is mournful to record that Hannibal was to poison himself in Bithynia, whither he had fled to escape the implacable pursuit launched against him by his life-long enemy, Rome.

XV

Legion vs Phalanx—The Battle of Pydna

Having made some cursory remarks on the effectiveness of the legionary system of the Romans as opposed to the Hellenistic or Macedonian-type phalanx, it might be both instructive and interesting if we were to examine in a little more detail one of the great battles where the two were in opposition, both from the historical and the wargaming points of view. We shall do this by once again directing our attention back to the Eastern Mediterranean and briefly relate the events leading up to the engagement we are to consider, which is in fact the Battle of Pydna, fought in Greece not far from Mount Olympus in 168 BC.

By the beginning of the second century BC Rome, freed from the incubus of the Second Punic War, was forced to turn her attention eastwards towards the continually warring and incessantly changing Hellenistic Kingdoms of Greece proper and of Asia Minor. It would be of little profit to trace the complex changes which had occurred in this political melting-pot during the previous century or so, and indeed space precludes all but our noting the salient points which might enable us to sketch out the necessary background. Macedon (now of course, but a remnant of the old Alexandrian empire) had pushed her area of influence across the Aegean Sea to the coasts of Asia Minor—Ionia—this being carried out in concert with the still powerful Seleucid Kingdom. The idea that their destiny should be dictated by a sort of Macedonian absentee landlord was not pleasing to the small states which were the first to feel the impact of this expansion and two, the mainland state of Pergamum and the island of Rhodes, appealed to Rome for protection, a proceeding which might suggest that already Rome was not without some reputation in the area. The Roman decision to aid the supplicants was probably influenced by the fact that Philip of Macedon, the King, had formerly made an alliance with Hannibal, conduct calculated to damn him in the eyes of the Carthaginian's pitiless foe. A force built around two legions was sent to Greece and promptly defeated the Macedonians at Cynocephalae in 197 BC but, following this, the Romans merely assured themselves of a permanent place in Greek councils and took no steps towards making a long-standing physical establishment in the country—or at least not for the time being.

Philip's ally, the Seleucid King Antiochus, now took up the cudgels and landed an army in Greece in 192 but he had as little success against the legions as his ally, first being driven from the country and then soundly beaten at Magnesia by a powerful Roman force under the famous Scipio Africanus. This battle had everything one might have expected it to have had—elephants, chariots and cataphracts, the last being the generic name for very heavy cavalry, with both horse and man at least partially armoured. In a favourable situation they could be very effective but, being slow and unmanoeuvrable, were vulnerable to an attack from flank or rear by lighter troops.

For some years there was an uncertain lull in the Eastern Mediterranean, broken only by small wars here and there, but the settlement imposed by Rome after Magnesia was unstable and it was apparent that soon there would have to be another major military intervention. The principal figure in the anti-Roman interest was Perseus, son of Philip and now King of Macedon, who seems to have had more than his share of devious not to say Machiavellian traits. He first attempted, under the guise of friendship with Rome, to undermine her authority by creating a kind of subterranean anti-Roman coalition, and in consequence of this he was appealed to by a number of Greek states to act as their saviour or protector. At this point the situation was immediately very much in his favour both financially and militarily, and it can be assumed that, had he acted with promptitude and decision, he might well have initiated a general rising against the pervasive and intangible presence of Rome. Of course, had this indeed taken place, how long such success would have lasted with the entire immense resources of Rome thrown into the struggle is very questionable. In the event, he considered it safer to remain on the defensive and in due course a Roman army of two legions, plus a large number of allied troops, sailed for Greece in 171 BC. Part of the army marched into Thessaly where it was met and defeated by Perseus who, alas, failed utterly to take any advantage from this initial and somewhat surprising success and remained totally passive. He seems indeed to be one of those characters who, having first created a situation, is rather taken aback by its implications and shies away from the Frankenstein-type monster he has built.

In any event, after a couple of changes in the leadership of the Roman army together with a few rather *opera-bouffe* manoeuvres and actions, another new commander arrived from Rome to take over the command. This was Lucius Aemilius Paullus, who was to take a firm grasp of the untidy and unsatisfactory situation in Greece. He seems to have been a competent soldier and quickly decided, having examined the situation, that it would be quite impractical to make a direct attack on King Perseus, who was established in an entrenched camp on the north side of the river Alpheus (which ran into the Gulf of Thessalonika). It was very strong, being well furnished with engines, the 'catapults' and so on which formed the heavy artillery of the day. He decided to use craft to force the Macedonians out of their near-fortress, by dint of attempting to persuade them that he was about to attack frontally and at the same time sending a strong force on a wide flanking march to approach the enemy from an unexpected direction. This plan, although it was discovered before it was well under way, was successful in that part of the army sent by Perseus to counter the manoeuvre was defeated, and in the circumstances the Macedonian King had no option but to fall back. He moved northwards, to take up a position to the southwest of the little town of Katerini (itself some ten miles south of Pydna which gave its name to the battle to follow). Both armies drew up on opposite sides of a little river, the Leucas. The time was late June so this must have been little more than a rivulet, forming as little obstacle to troop movement as did the Asopus at the Battle of Plataea.

The map shows the essential facts we have to know concerning the two opposing armies, although much indeed is only speculation, only the most fragmentary contemporary evidence being available but, having regard to the orthodox drawing up of Roman and Macedonian armies at this period, the layout is probably correct. Actually what we are mainly concerned with in the present context are the relative positions of the Roman legions and the Macedonian phalanx, and it would seem that they more or less faced each other. There is no

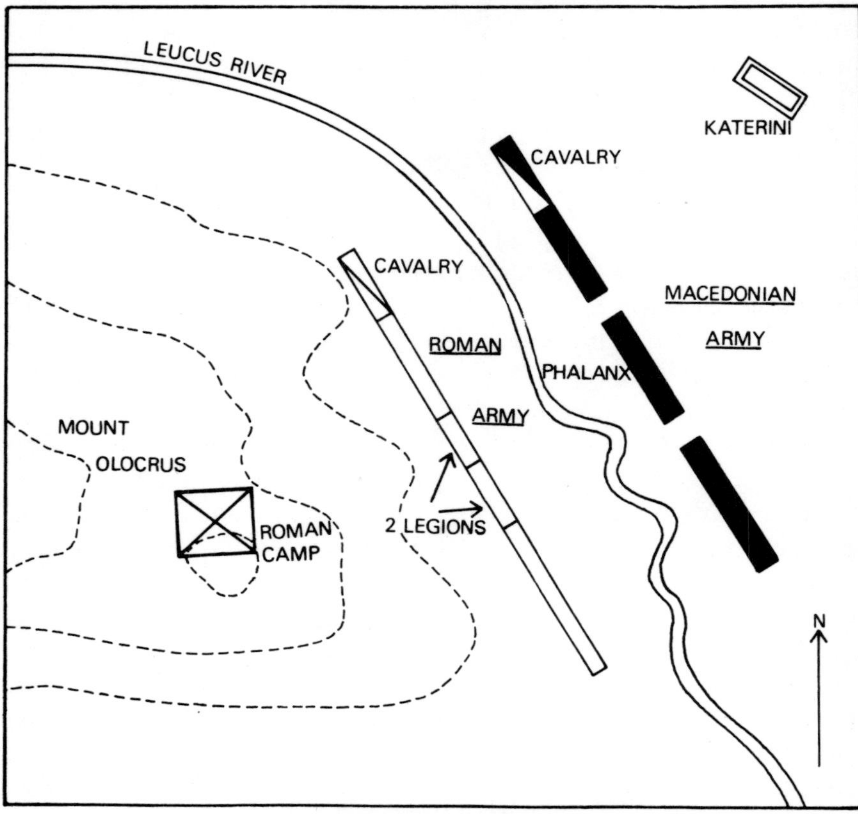

Elements of the Battle of Pydna 168 B C

evidence on the numerical strength of either army, but the legions can be estimated at something like 3,000 men each, although the establishment would obviously be affected by normal campaign wastage, sickness and straggling (even Rome being prone to this phenomenon). These totals are speculative, then, as is the suggestion that the phalanx might be about 10,000—it could indeed be very much more.

Strangely enough, it seems likely that the battle started by accident, with a straying Roman horse being pursued towards the Macedonian lines by its owners, who were challenged by some Thracians in Macedonian service, one of whom was slain, and with more and more men coming to join the fracas from both sides, Perseus and Aemilius had no alternative but to lead out their forces en masse and prepare for battle, forming their troops up as shown on the map. The Romans arrayed themselves on the lowest slopes of Mount Olocrus with a perfectly flat plain in front, but with more uneven ground to their rear. This was important, as we well know that a smooth terrain was essential for the movement and successful operation of the phalanx.

As far as we can judge, the Macedonian army was first to advance and, crossing the stream towards the Romans, they were attacked by the latter's right wing which, however, had the worst of the fight and was flung back towards the rough ground which we have mentioned. To conform with this involuntary movement the entire Roman line fell back, drawing with it the Macedonian centre, the great phalanx. The appearance of this mass of men at this first onfall, and the myriad gleaming sarissa points projecting from their ranks daunted the Romans considerably but under the firm hand of Aemilius they stood fast. Their steadiness was quickly increased by the sight of the phalanx beginning to show the first signs of disorder as it moved on to more uneven ground, gaps appearing here and there in what should have been an unbroken front, everything pointing to imminent disruption. With great promptitude the Romans poured into these intervals and, spreading to right and left, attacked the poorly protected flanks and rear of the phalangites. At the same time the Roman general led forward one of his legions into a major interval which had now opened between the Macedonian left wing and centre and fell upon the flank of the phalanx with gusto. The whole Roman army followed up and in the numerous separate actions the phalanx began to crack, then broke and fled, led indeed by the defeated Perseus and his relatively unscathed cavalry.

The sequel does not really concern us, but what does is that Pydna really marks the end of the era of the phalanx, although it did continue to appear in Hellenistic warfare for another fifty or sixty years but without great success and, as under Perseus, it was never exploited to its full capacity. As has been already pointed out, it did require a general of more than average skill, which Perseus certainly was not.

Pydna can be a most interesting combat in war game terms and is a battle which could well be chosen to illustrate the effectiveness of one's rules, especially in regard to the power of the phalanx against the legion and vice versa, and also, as a corollary, to compare the phalangite with his long sarissa—at this time of the 21 foot variety, longer than the old Alexandrian one—with the legionary. One can take two of the individual soldiers involved, Roman with pilum, shield and short sword, and phalangite with sarissa, possibly a short sword and virtually worthless round shield or buckler and match them point by point. Even should one accept that the later phalangite was equipped with bronze body armour he is still at a considerable disadvantage. Let us imagine they are approaching each other with a view to doing battle; at twenty yards or less the Roman hurls his heavy pilum, the recipient, the phalangite, having levelled his sarissa has, perforce, swung his small shield round on to his shoulders, and is consequently unarmoured (or, if the later type, has only his bronze corslet as protection, insufficient against the pilum). With the phalangite now possibly wounded, the legionary drives home with his sword. Multiply this proceeding by the number of men in the two formations and it will be seen that the phalanx, even before contact, has suffered very severely. The front ranks are in confusion and the legion is cutting its way into its vitals, thrusting aside any still levelled sarissae with their shields and stabbing at the holders with their deadly swords. Of course, at this really close range the Roman has practically no opposition, his close combat training being infinitely superior to that of his opponent. Obviously, once the legionaries are in its midst, the days of the phalanx are numbered, and it would seem that ordinarily there would be no doubt as to the eventual winner in such a conflict, once the overall protection of the wall of sarissae had been neutralised, as it was at Pydna.

In war game man-to-man terms we have to put a value on the impact of the legionary, that is, on the combined effect of pilum and sword, as opposed to the clumsy sarissa, as it certainly was once cohesion had been broken. The effectiveness ratio of legionary to phalangite might be reckoned as being 5 to 2, shall we say, but there are other considerations. It is but fair to give some advantage of weapon length to the sarissa, so we can add one to its value, thus producing a ratio now of 5 to 3. The extreme length of the weapon does make it possible for many more men to be effective so, in war game terms, it can fight on possibly four times the number of ranks available to a legionary formation. On the table a phalanx can fight four ranks deep as against the $1\frac{1}{2}$ ranks of the legion. A further and most important factor is that when fighting, the legionary has the full use of his shield while the phalangite has not, thus adding an additional 1 or 2 to the legionary value depending whether his phalangite opponent is armoured or otherwise, as in the case of the earlier Macedonian phalanx. All these of course are but the basic values and do not take into account any variable as provided by a dice throw.

The classic tactic for a wargamer faced with a phalanx is therefore to disorder it or reduce it to some confusion by missile fire, or by assailing it with javelin-throwing light infantry, whose proximity would inevitably cause the phalanx to level sarissae, leaving its members shieldless so that much more damage could be inflicted before a charge was made by heavy infantry or whatever. It would hardly seem necessary to point out that cavalry should never make any attempt to charge a phalanx frontally—there is no surer way to war game suicide.

Many are the occasions when phalanxes of various kinds are deployed on the war games table, and certainly its most successful users emulate, if they can, the tactics of Alexander by employing it to 'pin' an enemy, or even just to cause him some apprehension, and let no one claim that in a mere war game a player will never be nervous or tense. If not blood, and maybe not even tears, certainly much sweat has been expended around a table, which I can say from personal observation *and* experience.

XVI
Rome and the Barbarians

In the space available only brief reference can be made to the series of wars which later Imperial Rome waged against the multitude of nations and peoples lying beyond her sprawling frontier. Furthermore, the use of the word 'barbarians' in the chapter heading is in many cases quite inaccurate, some of the peoples involved having reached a degree of culture and civilisation quite equal to that of Rome. The term is really employed in the Greek sense and I have taken it—somewhat apologetically—to describe any race not yet assimilated by the Roman state and its traditional life-style. As such, there is a tremendous variation to be found in these nations. They can be as different as, for instance, the Picts and Britons from these islands, Dacians from north of the Danube, Germans from the dark Teutonic forests, and Sarmatians and Parthians from the other extremity of the Empire. For many hundreds of years warfare was almost continual and it is to the credit of the Roman army that so many different enemies and variegated tactical systems were met and successfully countered—if not immediately, at least usually in the long run. We shall make an attempt, if we may, to give just some indication of the manner in which the legions coped with the diverse military problems confronting them.

However, we must note that the legion is not exactly that to which we have already referred, for in the early years of the first century BC certain important changes had been made under the influence of one Gaius Marius, and these 'Marian Reforms' established the shape of the legion—the hard core of the army—for several hundred years. Its internal organisation was first of all largely changed, the three lines of hastati, princeps and triarii, with their different weapons and manipular organisation, giving way to a new tactical unit, the cohort, which was much larger than the maniple and numbered, with one exception, some 480 men. Each legion included ten such cohorts of this strength, except for the senior which had 800 men—a kind of double cohort. Each cohort was subdivided into 'centuries' of about 80 men, and to each legion was apportioned some 130 cavalry. Thus, on paper at least, the establishment rose to something between 5,000 and 6,000, but naturally this was very much an optimum and in the field I should think that any Roman general would have been happy to have his legions with a strength of 4,000 men or thereabouts. Weaponry had also changed; the long thrusting spear of the princeps and triarii being abolished, every legionary being now armed with the pilum, the heavy throwing javelin, formerly carried only by the hastati. Thus, the tactics of the entire legion in fact were now those of the old first line—the advance, the volley of pila and the rush to close quarters with shield and sword. One important psychological innovation was the granting of an 'eagle' to each legion, this device becoming, as every schoolboy knows, synonymous with all aspects of the power and majesty of Rome.

As time went on cavalry became rather more important in the scheme of things and many different types were developed as the centuries rolled by. Initially they were the customary light troops, mounted javelinmen, who usually if not always acted as skirmishers, and in the eastern portion of the Empire in the second century AD were to be found horse archers, although they were not seen frequently in the west. In contrast to the very light mounted archers were the super heavies we have already spoken of, the cataphracts. These in fact were not new and had existed since about the fifth century BC, the Massagetae, a people dwelling east of the Caspian Sea having, even at this early date, equipped their mounts with a bronze chestplate. They were probably able to do so only because they had bred a much heavier horse than was the norm for the period, and this was copied by several neighbouring nations notably the Persians.

From this simple beginning of an equine cuirass there developed a more or less complete horse armour, sometimes of leather sometimes of metal, this being found in the cavalry of many eastern nations, the Parthians for example, and more surprisingly for a mountain state, of the Armenians. In the Roman army these cavalry units were recruited from alien or subject peoples and it was in fact the Sarmatians who provided the cataphracts for the Roman service in the second century AD. The armament was a 12 foot spear called the 'kontos' plus a heavy sword, and with this weight of metal, on big horses, a Sarmatian cavalry charge was something which must have required a deal of stopping. The Parthians, whom we have just mentioned, were a very horsy-minded nation. Their territories lay about the Euphrates and their army was virtually wholly composed of two types of cavalry, light horse archers and cataphracts. The Parthians chalked up more

Later Roman mounted troops; heavy cavalry and Sarmatian cataphracts with the dreaded 'kontos'

than one success against Rome, particularly the Battle of Carrhae in 53 BC, when the Roman general Crassus advanced into Parthia with about 40,000 men, of whom 28,000 were legionaries, and was encountered by a Parthian force of about a quarter of his strength, a thousand cataphracts and ten thousand mounted archers. Parthian tactics were simple but prodigiously effective. The horse archers rode up as close as they could and sent volley after volley of arrows into the Roman ranks (the Parthian general, Surenas, with a fine eye for logistics, had organised a supply train of camels laden with a practically unlimited supply of arrows), and when, galled into activity, the Romans advanced from the enormous square they had formed, the cataphracts replaced the horse archers and drove them back. Their all-round horse armour was so impenetrable that all the Romans could do was to lie on the ground and stab upwards into the unprotected bellies of the horses as they passed. Ultimately the Roman army began a disastrous retreat in which only a quarter of the entire force made good its escape, Crassus himself being killed and the remainder suffering similarly or being taken prisoner.

Only a few years earlier there had occurred a very fine run thing at the other end of the Empire—I use the term with extreme looseness, the Empire being not yet established, but it is a good, evocative word—this incident involving no less a personage than the mighty Julius Caesar, in the process of making his name as a general by dint of years of successful campaigning against the Gauls. He is another we have to acknowledge as one of the Great Captains and this battle, that of the Sambre, fought in 57 BC, is indeed a classic instance of how one man converted an imminent defeat into a victory. As was the custom, a Roman army of six legions under Caesar was preparing camp for the night on some high ground looking down on the River Sambre—it was Roman practice at every halt to ensconce themselves in a very formal camp with ditches, stockades and so on—when from the dense forests on the other side of the river there burst three huge Gallic warbands, a tribal confederation of Nervii, Atrebates and Viromandui, all extremely pugnacious and ferocious tribes. They rapidly crossed the Sambre and came racing in their thousands up the slope towards the completely surprised Romans who were going

Asiatic archers were extensively used in the eastern portions of the Roman domains

about their lawful occasions, digging ditches, laying out the camp and so on. For once there had been no scouting, or what there had been had proved ineffective, and the Romans were caught in an advanced state of disorganisation. Still, splendid fighting men that they were, they clubbed together in ad hoc units and with Caesar moving from group to group encouraging and inspiring them, they held the enemy at bay. Still, fighting was desperate in the extreme and there seems to be no question that, had Caesar not been present, even Roman discipline and training would have been insufficient to prevent the legions disintegrating. Thanks to his presence, however, they held on and finally the arrival of two other legions, which had been marching more slowly than the others, turned the day in favour of the Romans, the Gauls being defeated with great slaughter. The Battle of the Sambre is a most interesting encounter and as a war game it has much to recommend it—as we shall see ere long.

Incidentally, the tactics of the Gauls were not dissimilar to those of their near neighbours, the Germans. They had little cavalry, the bulk of their warbands consisting of fast-moving infantry armed with javelins and long swords. The latter had no points and were used only for slashing, but the metal from which they were made was not of the best and the blades tended to bend on impact. Armed with these lengthy weapons warriors had to fight in fairly open order, to allow the swordsman sufficient room in which to operate. One might imagine that in close combat with the more closely ranked Romans two at least of the latter would be available to take on one Gaul or German. By definition, these western tribes might be extended to include the Britons, probably the last people to use the chariot in war. In his writings, Caesar speaks of their skill in handling their vehicles which were, in fact, of extremely light construction with spoked wheels, not the solid blocks of wood one occasionally sees depicted by the uninformed. Their tactics did not usually employ the charge pressed home, their preference being for the driver to bring the chariot at full speed up to an enemy, then to wheel away while his crewmen hurled javelins. This individual did occasionally dismount to fight at close quarters, but was always prepared to leap back into the waiting chariot should circumstances advise such a course.

As well as reverses in the east, the Romans also had their setbacks in the west, for example in 9 AD when three legions were destroyed in several days' operations in the forests of the Teutoburger Wald by the German Arminius, once an ally of Rome. In Britain the rough handling of the Ninth Legion by Boudicca during her rebellion in 60–61 AD is another example, but such happenings almost always had some element of the ambuscade in the situation and in the open field, in a set piece battle, the discipline and tactics of the legions were generally more than a match for the undisciplined, all-out charge of the barbarian tribes with whom they fought. This of course was the favourite, possibly the only tactic of the barbarian—a headlong rush, if possible downhill, in a great screaming mass, highly frightening but, once held, relatively easy to defeat.

Of course, leaving aside barbarians for a moment, we must not forget that Rome was frequently rent by internal strife and the Civil War, which lasted from 50 to 48 BC and ended with the Battle of Pharsalus between Caesar and Pompey, gives ample opportunity for the wargamer to match legion against legion. From personal experience I can confirm that when two such units meet the resulting crash is of sufficient magnitude to be audible halfway along the Appian Way! To be honest, though, it must be admitted that a Roman versus Roman war game is usually a pretty gritty sort of thing with not a great deal of movement or manoeuvre, for to get a result sooner or later the legionary infantry must meet and then it is a case of prolonged slogging until one side gives way. Such battles too are to be found in the fighting following the death of Caesar—the Battle of Philippi in 42 BC, for example, was a massive affair with no less than seventeen legions being engaged on each side.

Once, though, that a definite ruler had emerged in the person of Octavian, Caesar's heir, the legion was well on its way to becoming the most formidable fighting unit of its time, for, though initially there seemed to be stability throughout the Mediterranean it was largely ephemeral, and in the years to follow there was continual battering at the frontiers by all sorts of barbarian nations. This was primarily due to the immense pressures arising from the great movements of peoples in central Asia. Octavian—or Augustus, as he became—created a regular and permanent army of 25 legions, prior to this troops having been raised and disbanded as the need arose and the emergency receded. In addition to the legions there was a vast number of auxiliary units, many still recruited from conquered peoples.

Batavians, Gauls, Germans and Spaniards were among those formed into such units, some being infantry—javelinmen, slingers and archers—but a high proportion was of mounted men, each legion being accompanied by its associated cavalry, in certain cases this being one unit of light and one of partially armoured troopers.

During the first century AD it was becoming customary for each of the legions proper to have its own permanent base, which formed a kind of depot, wherein would remain a small garrison while the bulk of the legion went off on active service. All this time more and more states were being incorporated into the Roman Empire by one means or another, particularly in Asia Minor, and there was unceasing warfare with the tribes along the Danube, the Dacians especially. No less than ten legions were required before this fierce people could be finally subjugated, their territory for many years forming Rome's only holding across the Danube. Even with the great resources disposable the pressure on the Roman Army increased steadily, and more and more races and peoples were on the move. Among them were the Goths, who originated in Southern Russia, the Alans and the Vandals, while the tribes beyond the Rhine were also beginning to associate themselves into greater confederations—Alemanni and Franks. Far to the east the Parthians had been overthrown by the vigorous Sassanid Persians, who began a strong eastwards threat against the Roman client kingdom of Armenia, their aim being to restore the ancient glories of the Achaemenids. So much pressure was being brought against the 'limes'—the frontier—that by 230 AD no less than 33 legions were deployed along the imperial border, which stretched from northern England, across Gaul and along the Rhine to the Danube, and down the eastern marches of Asia Minor to Palestine and Egypt. The resulting strain was beginning to be felt most acutely, and not the least important feature in the chain of events was the gradual eroding of the iron discipline of the Roman legionary. The system of permanent garrisons dissociated the troops from the central authority and this was reflected in numerous local mutinies—some very widespread and serious—and the occasional proclamations of local generals as emperors in various parts of the Roman-held territories.

The middle of the third century AD saw a succession of disasters and the Empire was sorely rocked by the impact of the Goths in the Balkans, and by the raids of the Alemanni and Franks into Gaul. The

Sassanids reconquered Armenia and in 260 AD actually defeated and captured the Emperor Valerian. The confusion and virtual anarchy prevailing throughout the empire about this time is reflected in there being at one moment no less than three claimants for the imperial title. As the century drew to a close there came a kind of revival but although Aurelian managed to restore some degree of stability to the frontier, under Diocletian in 286 AD the empire was divided, as it was becoming increasingly more apparent that the length of frontier and the problems of communication were too much for one central authority to control. A 'co-Emperor' was appointed to govern the West, and still later a 'Caesar' for each 'Augustus' was created to act mainly as military commanders for East and West respectively.

It would not be to our advantage to pursue our story further; it would occupy far too much space and time and certainly could not be adequately done in the permitted framework of this book. Only the briefest of references can be made to Roman military organisation during the centuries we have so summarily covered. We have mentioned the establishment of bases for the legions in the second century and that there was a considerable increase in the number of supporting troops, particularly those actually attached to the legions, although all these auxiliaries were by no means of the highest quality. In the third century AD—and this must be our final period—we have noted the many changes in organisation, although military affairs were generally in a state of flux. Usually the complete legion was not employed as such, but a 'vexilation' or large detachment was used as the basis of a task force, this being made into a sort of battle group by the attachment of cohorts of light infantry, together with units of cavalry. Towards the end of the century some of the cavalry would be armoured cataphracts, generally of Sarmatian origin, whom we have spoken of already. Indeed as time went on there was a startling increase in the number of 'barbarian' units in the Roman army, Aurelian for instance even creating auxiliary troops from the former menaces, the Franks and Alemanni, while some emperors had a completely Germanic bodyguard. At the time of the division of the Empire into East and West the strength of the army must have been at its peak, no less than 60 legions being 'on the strength', to employ a rather later term, although it may be questionable how many of them ever reached the prescribed establishment, being possibly in some cases little more than 'paper' units only.

Traditional legionary tactics—those of pilum and sword—were slowly becoming obsolete as we approach the end of the story. Long thrusting spears, similar to those of the old princeps and triarii were now used, as they were found to be more effective against the increasing numbers of heavy cavalry which were being deployed and the mounted arm as a whole was achieving a more important role in the line of battle. The organisation of the Roman army at this time and later is a fascinating and complex study regrettably far beyond the scope of the present work. It remains for us to select for examination—as a war game—an occasion when the legions took the field in their classic form. It would also be a good thing were we to select a battle wherein in a manner of speaking we can kill two birds with one stone, ie. to study the clash of Roman and barbarian, and also to see how surprise can be operated with success in a war game. Both will be fulfilled in the re-enactment, wherein we shall see just how the tactical skill of the protagonists shape up to the pre-arranged situation.

XVII
The Battle of the Sambre as a War Game

The shrewd reader will have deduced from the previous chapter the battle destined to be our next refight. It was said to be of great interest historically and to be particularly noteworthy in that it would be possible to introduce the element of surprise, this being a feature which adds greatly to the interest and which tends to make a war game even more fascinating than the commonplace one where the contending forces are set out in full view of each other, often at opposite ends of the table. The engagement we are to deal with, then, is the Battle of the Sambre in 57 BC, where the Romans, under Julius Caesar, fought a great horde of Gallic barbarian warriors.

Having briefly outlined, in the previous chapter, the background to and the course of this quite unusual engagement, no more need be said about its historical side other than to give the strengths of the armies engaged prior to reducing them to war game terms. Basically the Roman army consisted of eight legions, six of them, immediately before the fighting, being engaged in settling into camp, while two further legions were marching up with the baggage train. Now, as always in such circumstances, a certain amount of thought is necessary to decide upon a reasonable number of actual miniature figures to represent the units engaged. In this case, such thinking was governed primarily by the map and the corresponding table area over which the first six legions were spread in their camping operation, the idea being that individual figures could be spread about fairly widely to represent the legionaries digging ditches, erecting palisades and so on. This dispersion was also calculated to give the attackers a reasonable chance of success. Furthermore, once a figure has been established for the Romans, it would be relatively simple to estimate the numbers of Gauls. These figures are important for after all there is a limit to the number of troops the average wargamer can deploy. In the event it seems that the Gauls were something like twice as numerous as the Romans. After some consideration of numbers and space, then, it was determined that a legion should be represented by 30 actual figures (strictly speaking, on the customary 20 to 1 ratio which is normal, this would make the equivalent of a cohort approximately, and not a legion, but let us not concern ourselves too much with this). Thus the eight legions of the Roman army would be 240 actual figures, not a bad array by any standards.

Now, the number of Gauls is naturally more difficult to ascertain than that of the Romans, whose strength can be deduced from the number serving in a legion. It has already been suggested that a reasonable total for a legion under field conditions would be about 4,000, so by this token the legionary portion of the Roman army would be, we hope, something in the order of 32,000 men. To this must be added, incidentally, a thousand or two Gallic allied cavalry, together with some slingers and archers, although these were few and far between and might safely be ignored. However, it was thought proper to include a few cavalry, for there is a specific mention of them in the histories, and it was decided that there should be a unit of 15 figures—Gallic auxiliary horsemen. Suppose then that overall Roman strength was 35,000 we find that, give or take a few thousands, we have to assume that the strength of their barbarian adversaries was virtually double this number. Strongest of these by far were the Nervii, who may have mustered as many as 60,000 warriors, with 10,000 Viromandui and 15,000 Atrebates, thus giving a grand total of 85,000 men. This may well be an inflated figure and we could with some justification reduce it to, say, 70,000. This is not without malice aforethought as it gives a straight two-to-one ratio vis-à-vis the Romans, but this little bit of juggling availed us naught as, when it came to figures, the Gauls numbered 400 men. This is not an accurate conversion but the plain fact was that this was the total number of figures available, so what else could one do? In the event they proved more than enough.

Again for convenience this total was apportioned rather more differently than was correct; the Nervii were 200 strong, and the Viromandui and Atrebates 100 each. Although historically each tribal confederation would have its own 'supremo', there would be

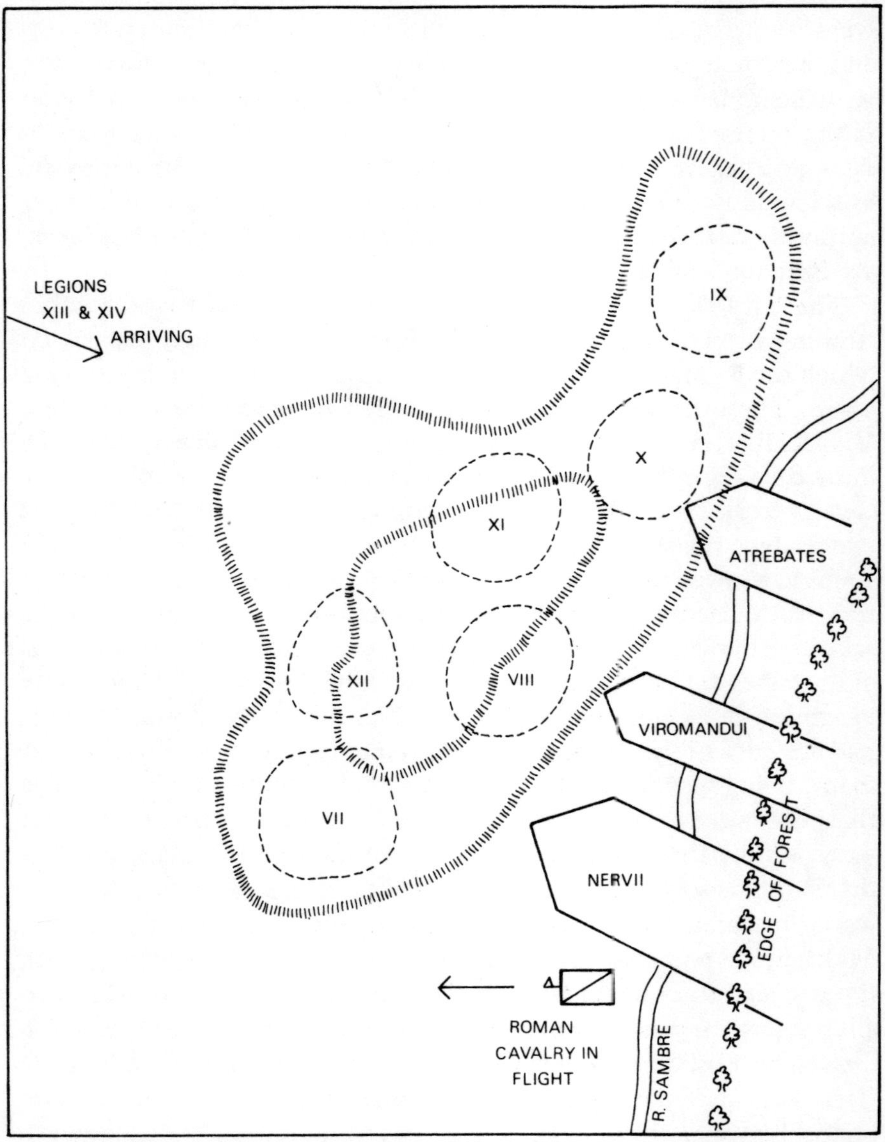

Map 1: the Battle of the Sambre as a war game. Diagram of a situation at the end of the first period, dotted circles show dispersal area of legions

much division into sub-tribes. This accords well with the ordinary system of command and—again quite arbitrarily—the sub-tribes were said to be of 50 figures each, thus the Nervii had four such units and the others two each. All of course were infantry, armed with long swords and javelins, plus perhaps a light shield. The chiefs and other senior tribesmen would certainly have some sort of armour but this does not affect the immediate issue. Here and there in the tribes would be a few slingers but they can be ignored—missile fire was unimportant in the circumstances. The only mounted troops on the field were the Romano-Gallic cavalry.

The battlefield is as shown on the map, with the River Sambre running across its eastern side. To the west is the high ground on which the Romans have just dispersed to gather material for making camp. The war game legions—numbered as were their prototypes, VII, XII, VIII, XI, X and IX—were well scattered about, with each individual figure several inches from its nearest neighbour. The cavalry had crossed the river but as it is highly unlikely that they had penetrated the forest their counterparts were stationed in the position shown (they could not really have gone into the forest where they could hardly have failed to see something of the Gauls). For the sake of achieving as complete a surprise as possible only the positions of the tribes has been noted on paper, and the Romans were ostensibly in ignorance of what was to happen. Neither of the Roman players had any idea of the battle which was to be re-fought but then both of them, being as wily as the next man, had doubtless a shrewd idea that they were to be under immediate attack, and were not without some inkling that it would emanate from the wooded country across the Sambre—obviously the most likely quarter. The two Romans were, in effect, Derek Casey—'Julius Caesar', no less—and David Matthews, who was in command of the two legions marching up with the baggage train towards the main army. Opposing them were two allies whose names have also been met with in these pages, Andrew Green, back in his customary role as the barbarian leader, Andra the Berserker, and Alan Stoneman, who shared the fortunes of the war-bands between them. The theoretical positions of the tribes are shown on the map.

The first period began with the appearance of the baggage train with the two legions—XIII and XIV—on the far side of the table, this immediately and historically being the signal for the attack, and from the forest came pouring forth the Gauls, warband after warband. First to be affected by this avalanche was the Romano-Gallic cavalry who, as soon as the enemy burst from the woods, had to have a 'first sighting test' of their morale. It was disastrous. Being a barbarian unit, ordinary 'wild' dice were used and, although a reasonable total was thrown, circumstances—the unfriendly cover, the huge numbers of enemy approaching, their lack of immediate support and the distance from them of their general—were such as to cause them simply to turn and flee in confusion. As for the Romans, despite their disorganisation, their morale, when tested, was satisfactory everywhere, although it was touch and go for more than one of the legions.

In any case, over the Sambre came the Gauls. The river, incidentally, does not seem to have offered much of an obstacle to them, but as it surely must have caused some delay, it had been decided previously that, although fordable everywhere, this would take a little time, and accordingly during the crossing the infantry move would be cut by half, and that of cavalry by three quarters. In spite of the latter the fleeing Gallic horse were able to get over without being caught by the enemy.

Roman reaction was automatic and predictable, each legion very properly beginning to close up to its 'eagle', but there was still little cohesion when on the third period the first of the Nervii warbands hit them—these having been closest to the edge of the forest. Fighting 'disorganised', the VIII and XII were pressed back, but XI and X stood fast against the Viromandui and indeed one of the two Atrebates warbands was broken by Legion IX and fled. By period 4 the entire Gallic army had crossed the river and all the tribes were flinging themselves at the Romans. In point of fact their very numbers were something of a hindrance, the line of attack having been directly up the slope against the Romans and there was no immediate attempt by the Nervii warband on the extreme left to make any flanking move round the Roman rear, nor to move rapidly forward and hold off the other two legions which were now several moves in from their arrival point and approaching fast, although it would be four or five periods before they could make their presence felt. Despite still being in some disarray the legionaries were beginning to assemble into fairly substantial bodies and succeeded in forming a more or less continuous line along the ridge. 'Caesar' himself had at the outset been with his favourite Legion X. This had been a totally fortuitous occurrence for as no evidence was available to state exactly where he had been at the moment of attack, an ordinary die was thrown, each face referring to one particular legion and as luck would have it, X was the one indicated.

By this time the small body of cavalry was practically out of sight, having failed to rally, and having passed the two reinforcement legions. In so doing, of course, the latter had to have a test of their own morale, this taking place always when a unit breaks or flees within a certain distance of another—panic as everyone knows is very readily communicable. This indeed had resulted in the two legions having to halt for two periods. This, although unfortunate for the Romans, strikes one as being a fairly realistic proceeding as, after all, they would doubtless be prey to some hesitation when passed by flying cavalry as well as seeing their comrades ahead being assailed by myriad enemy tribesmen—the word 'hostiles' comes unbidden to the mind! After the two stationary periods they went forward again, but their hesitation although not completely crucial was pretty significant. During it one Viromandui warband had first pushed back Legion VIII—still not completely reorganised with some men still to come up—then, with the aid of the 'follow-up' bonus, equivalent to that given for a charge, *and* with the added advantage of fighting downhill, as they had moved forward from the crest, and plus a good dice throw had broken the unfortunate legion which turned and fled, with the Gauls in hot pursuit. As luck would have it, another warband was directly behind the victorious one and this one, pouring into the interval, swung to its right and went for the flank and rear of Legion XI. The result was a foregone conclusion. Heavily engaged to its front, and now hit in flank and rear, it collapsed.

Morale tests now followed all round, naturally, and in spite of good dice throws both X and XI were obliged to carry out a retreat and the entire Roman army had no option but to fall back, 'Caesar' himself—like Napoleon a great many years later at Waterloo—seeking refuge within the ranks of his Old Guard, in this case Legion X. Legions XIII and XIV were still not close enough to influence matters and indeed had to halt and form a rallying point for the others. Actually it seemed that the tribesmen, having quite definitely established that they had won a striking victory, preferred to halt and plunder the Roman camp, or what there was of it, this being deemed highly proper and traditional.

The Battle of the Sambre: the Romans are dispersed all along the Ridge and in the foreground their allied cavalry is in flight. The leading elements of the Nervii are crossing the river

This, our second refight of an historic battle, did not follow the pattern of its predecessor—Plataea—in having the same result as its prototype, but this is really what the war game is all about, is it not? To take history by the forelock, in a manner of speaking, and give it a vigorous shaking is highly satisfying to the wargamers who have done just that, and Messrs Green and Stoneman were well pleased with the result, as well they might be.

Would it have been proper to give greater value to the Romans because of the presence of Caesar? A doubtful thing, this, as it would seem proper that the value given to a normal commanding general would be sufficient—after all, he did blunder more than a little in his reconnaissance before the battle. What might have been more appropriate would have been for the preparer of the scenario, myself in fact, to have altered the original setup to allow for the Romans' placing pickets along the river, or indeed one of the legions. This might have held up the Gauls sufficiently to allow the Romans time to get together, but this is wisdom after the event, and in any case the aim was to reproduce as accurately as possible the initial position of the actual battle. Nevertheless the game was an exciting one, which is really the raison d'etre of the whole operation.

XVIII
Conclusion

And so we reach the end of our long journey through history. It has taken us from the Egypt of the Pharaohs, through Persia, Greece and Gaul, to Imperial Rome and if only a little of the fascination, the colour and—dare I say?—the glamour of this most tremendous period has been communicated to the reader, then my objective will most certainly have been achieved. I cannot think indeed that very many voices could be raised in protest against a claim that for the purposes of historical study as well as for the practical demands of wargaming, there can be no better period than that whose salient points an attempt has been made to cover. Of course it is difficult to pinpoint exactly where the study of history leaves off and the practice of wargaming begins; each is so irrevocably bound up with the other. Indeed it has long been a contention of mine that, in a way, the latter is simply an end product of the former. Great numbers and types of people are profoundly interested in the past and more and more is wargaming beginning to be recognized as one of the most graphic and exciting ways of reconstructing some of the striking events of bygone years and, in these enlightened days, many schools, with official blessing, have their war games clubs and societies. This is unbelievably rewarding to one who well remembers when the wargamer was indeed a *rara avis* and was fain to blush with embarrassment when quizzed about his hobby by the sceptical or the unbeliever. Few such Philistines are with us now, and only the rarest of individuals continue to regard the wargamer as a psychopath or retarded child.

In all this proliferation of wargamers and their societies it is inevitable that such a dramatic cross-section of the population, as is supplied by the brotherhood of players, should have healthy differences of opinion on many aspects of wargaming, these being mostly marked in connection with the submission of rules. There is—I go so far as to say, happily—no national body or group to lay down for all and sundry just how a game should be conducted or that it should be conducted according to some specific and arbitrary code of rules. That this—the almost general acceptance of one set of rules—should operate in the case of the ancient period is indeed coincidental, but were it not, it would be a good thing for people to have their own ideas and to put them forward in the proper place and at the proper time with a view to creating a general improvement of the quality and realism of war games in general. It is for this reason that I am profoundly reluctant to be in any way dogmatic in these pages. To do so is an invidious practice and what I have written in the way of suggestions are only intended as such, and although I and the great majority of 'ancients' subscribe, as it happens, to one particular body of rules, there is no compulsion upon them to do so—the choice is their own, either to adopt someone else's rules or to compile their own.

Once having done this, of course, it is of the essence to ensure, once battle has been joined, that the game flows smoothly and without any break, whether it be for argument, for disagreement or whatever. Wargamers are as individual as the next man, and the competitive spirit is high—it is in fact a very cogent reason for playing a war game —and the time will come when it will be necessary to seek 'counsel's opinion' for the interpretation of this or that situation. In connection with this, over many years of active wargaming I can hardly recall one engagement wherein there arose no combination of circumstance or where there developed no novel or unusual position requiring study and clarification, just to determine what would really have taken place historically. This is where the services of an umpire both unbiased and knowledgeable become absolutely necessary. It is not always possible to have one but in the group of which I am a member the system works most efficiently and amicably.

We use one large table and, despite on an average playing once a week, it does mean, with a dozen or more people in the organisation—albeit a very loose one—that most of the time several are present only in the capacity of spectators. Happily, all are most enthusiastic and while naturally preferring to play are nevertheless ready and willing to take part in a supervisory role. First, there are the two players, of course, then we have an umpire, who will ordinarily be one of the most experienced of the group. He might well be assisted by another, who could be referred to as the 'measurer', for want of a better word. Armed with steel tape, he carries out the necessary measurements, particularly when ranges and morale are being assessed—such things as 'how many units are within support distance', in the case of friends, or the reverse in the case of the enemy—another player will be occupied with making the total of 'pluses' or 'minuses' on a scratch pad, and will then announce the result, having compared his mathematics with the appropriate chart in the rules. If the battle is a large one, with many troops involved, then others will pitch in to assist in moving the figures. All this creates a feeling of involvement and in all modesty I can claim that our 'system' is as smoothly flowing as any I have experienced or witnessed. Quite apart from all this by the way, it has been stated by various spectators, for instance at conventions where all sorts of different periods of game are taking place, that the 'ancient' war game table is always the best behaved! I should personally be reluctant to make such a claim myself, but there it is—it could be true.

Be that as it may, what really is correct is that without doubt in recent years the increase in ancient wargaming has been truly phenomenal and more and more people are being drawn to it every day. The causes of this are many and complex and are not unconnected with the emergence of good rules and the decision of several model manufacturers to produce figures suitable for wargaming in the ancient period. Time was when little was available for wargaming other than in the 'horse and musket' era and the modern, but now, as a glance at the appropriate appendix will reveal, this is now far from being the case. Everything is now going for the ancient wargamer, and if this book has done only a little to engender interest in this most thrilling of periods and has provided possibly just a little in the way of interesting reading for the player already committed then its purpose will have been very adequately fulfilled.

Appendix I— The Single Figure War Game

One increasingly popular facet of present day wargaming and one which seems to fall very readily into the ancient period is the 'single figure' game. This is not actually the best of descriptions, being not necessarily entirely accurate, but what is meant is the sort of game which does not require the employment of masses of figures, but actually only one per side or possibly half a dozen or so. This has the immediately apparent advantage of requiring a great deal less in the way of playing area, and of course less expense or trouble in purchasing or constructing a large war game army, when there is also the risk of becoming prey to a common malady—a *rage de nombre*—where the sufferer goes to enormous lengths in accumulating masses of troops which may rest in boxes or on shelves for lengthy periods as they await painting. In this single figure game one can also venture into the realm of figures larger than the customary 25mm type and cut something of a dash with the 54mm showcase sort of model.

One such type of game is the medieval, which in effect can be either the reproduction of a tourney with mounted knights or a small action with a baron and a few sergeants-at-arms on each side, but what I think is particularly interesting in the context of our ancient period is the Gladiatorial Combat. We shall have to look upon this with a certain degree of humour, because historically the spectacle in the Roman arena, starting with straightforward man-to-man fights (to the death it must be said and bloody withal), completely degenerated into quite ghastly bloodbaths wherein thousands of men, women and children were sacrificed at the instigation of the Emperors and their familiars to provide entertainment for the populace. However, these bad old days are long over and considerable enjoyment can be had, in a completely inoffensive way, by organising a combat between two miniature gladiators on a small table, which indeed need not be larger than an ordinary chessboard.

Rules for such games have been published—I know of two such—and they do provide for the occasion where neither the time nor the space allows the setting out of a major sort of war game. They do allow for a certain amount of acumen, not to say guile, on the part of the players, of whom there will usually be two, although groups can also stage a kind of 'Roman Games' if it be so desired.

Single figure wargaming. Two noble Romans watch a trial of strength between a 'Thracian' and a 'Retiarius'

Virtually all the colourful types of combatants that strutted their little hour in the arena can be had in miniature from the 'retiarius' with trident and net (who has not seen films with such characters, *Spartacus* and *Demetrius and the Gladiators* come to mind) to the heavily armed 'Thracian', with cuirass, heavy helmet and metal greaves, using a curved sort of scimitar, and the 'velite', a much more lightly equipped individual with small round shield, short sword and spear, but little in the way of armour. Actually, the history of the gladiatorial combat is an enthralling one and much has been written about it, but there is far too much variety of it for us to discuss here.

The equipment required is really minimal—only a couple of gladiatorial figures, a board 'gridded' in squares, 1½ to 2 inches to the side, dice, pencil and paper and one is in business. Once the 'Entrance of the Gladiators' has been played and the fighters are in position at opposite sides of the board, then the players write 'orders' or more properly 'intentions' for the move, or period. Initially these would cover the approach to each other, but once in close juxtaposition one to the other, each player notes the actual stroke he intends to make and that part of the other's 'corpus' against which he hopes to make it. The retiarius of course has a longer range with his net than say the Thracian with his short sword. The latter would have to be in a square adjacent to one occupied by his rival before he could deliver a blow, while the retiarius would be able to 'cast' from possibly three squares away, although while so doing he could not move, and vice versa.

Charts are provided to establish the effectiveness of the blows struck—if they are not 'misses' indeed— and the result will be judged by comparison of dice throw and chart. Let it not be thought that only luck is the governing feature in the combat, for the clever player will either dodge an expected blow, or move in the most judicious manner to approach his opponent from flank or rear, and if he can, as I have seen done, sum up his enemy's intentions and state of mind, he is well on the way to victory. The effect of the blows struck will not inevitably be fatal, wounds being allowed for and they will have the effect of slowing up their unfortunate recipient and rendering him more liable to suffer a coup de grace.

The more Machiavellian sort of player need not be satisfied with being a human combatant but can insist that his enemy be a 'bestiarius' and himself appear in the guise of tiger or lion. In this case, though, his beast will be subject to certain natural laws and he will not be able completely to control it as he would in the case of a human gladiator.

This is all really part of the less serious side of the ancient war game and certainly, if nothing else, it gives the opportunity for having some very fine figures, several of which have been recently produced, both in the 54mm and 25mm scales. It is good and harmless fun, a salutary relaxation from the cares and vicissitudes of commanding a legionary army or the muster of a Greek city state.

Appendix II—Terrain

Broadly speaking, there are two types of terrain which can be used for wargaming, these being the dioramic (or pictorial) and the purely functional and it will be evident from the photographs in this book just which kind I prefer. This results from a long experience of various sorts of war games in which I have found that it is next to impossible satisfactorily to combine the two varieties to obtain a maximum of effect and utility. I should be the first to admit, naturally, that there is no question but that carefully prepared scenic pieces (made in the same way and with the same attention to detail as the static background for a model railway layout or the craftsman produced dioramic setting for a group of figures) are extremely pleasing to the eye but they have built-in disadvantages where wargaming is concerned.

Much of course will depend on the frequency of use to which some particular piece of terrain is put. In circumstances where war games are few and far between and every one something of a special event occurring once every other month or so, or when some extra effort is being made to stage a special battle, then it might be a good thing to set up an elaborately prepared terrain, with all the effects of bushes, streams, trees and so on that may be necessary. Where war games are more frequent, when for example a different game is fought every week and a different terrain is thereby required, this is not really practicable and some system must be devised to permit of quick changes and the use of pieces which can be readily transferred from one table to another. In this connection I am thinking particularly of hills and how they may be represented. The first requisite is that miniature troops may be able to stand securely thereon—nothing is more unpleasing than rows of figures lying prone and unlovely on a too steep slope and if the latter has any incline worth speaking of this will surely happen. The scheme I advocate is the use of thick insulation board or expanded polystyrene cut to shape to represent contours. One or more can be placed on top of each other to give the same effect as a series of contours in an ordnance survey map. The closer the contours the steeper the incline and movement upwards can be slowed down in accordance with the steepness of the slope.

Wooded areas can be shown quite easily by a ring of trees—either home-made or commercial products, both are effective—and river sections of plastic can also be purchased. The illustrations herein show them to be most successful.

However, it was a fairly general practice in ancient warfare (for reasons which I feel have been made plain) to seek ground where no drastic irregularities existed to hamper movement—far less use of cover being made than in later wars. Battles were, however, fought in the most unfavourable circumstances and in the densest of woods, for instance the Teutoburger Wald already mentioned, while of course there is the Battle of the Sambre we fought, showing how cover could be used to obtain tactical surprise. Basically, though, the question of terrain is strictly a matter for the individual but if one's war game table provides the necessary tactical features and they do not offend the eye overmuch then this should be sufficient—after all it is the sight of troops en masse which provides the aesthetic appeal and certainly in none of the photographs used in this book does the functional nature of the terrain detract from the visual effect.

Buildings also can be dispensed with—the latter day use of farms, buildings, and villages as strong points was simply not on when it came to ancient battles. There could be street fighting when a town was stormed or a village was surprised but the former is a highly specialised form of ancient wargaming requiring a great deal of time and while a raid on a village is indeed a valid war game the question of house-to-house fighting is a complicated one. Rules can be made to cover this contingency but they are not to everyone's taste and can well be left for those desirous of creating something different.

The influence exercised by terrain on the actual process of fighting can easily be established. We have mentioned in our account of the Battle of the Sambre how the river reduced cavalry movement by three quarters and infantry by a half. There can also be lesser rivers or easier fords which have no retarding effect and where the stream can be crossed without delay at any point. Moving up slopes could result in a slowing down or, if using the contour system spoken of, then any manoeuvre on contours above a certain height—say anything more than one contour high—would constitute moving on rough or broken ground, as the summit of a hill might well be, and manoeuvring would be inhibited as in the case of troops crossing a river.

Fighting or charging downhill would most justifiably be given an impetus advantage, or conversely troops advancing or fighting uphill could be penalised correspondingly—it is a matter of choice, and in this, freedom of selection applies, as it does to the whole question of the use of such features.

Appendix III—The 'Ancient' Rules of the War Games Research Group

The *War Games Rules, 1000 BC to 500 AD,* published by the War Games Research Group, first came to the notice of a grateful wargaming public, or at least the 'ancient' section thereof, in 1969, and the event was something of a milestone in wargaming history, not only that concerning the ancient period but in fact any other that could be named. The reason for this was that, for the first time, a book of rules based on sound research and considerable playing experience had been made generally available, and due to their high quality and to their being used for the ancient war games at successive National War Games Conventions, the result was that they were adopted by ancient wargamers throughout almost the entire country, to the exclusion of other rules. They are now in current employment not only all over the United Kingdom, but in the United States, Australia and New Zealand.

This, incidentally, is not to say that when the *Research Group Rules* were produced there were no others of merit in existence but those that were in use had never been brought to the notice of any large section of enthusiasts. It was also possibly somewhat fortuitous that about the same time more than one manufacturer of miniature figures had sensed the advantages of providing 25mm metal wargaming figures for the period and these were coming on to the market in considerable numbers. So, from one cause or another, not the least being the excellence of the *Rules* in conception and operation, they now govern the playing of ancient war games practically everywhere.

At first sight they are not uncomplicated, but once one has had the experience of conducting two or three games under their auspices, their principles become perfectly clear and they do provide a most smoothly flowing and realistic game. Indeed all the 'battles' detailed in these pages and indeed the majority of the principles I propound coincide almost entirely with those laid down by the Research Group.

What makes them especially noteworthy, this indeed being the case with any good rules of whatever the period, is that in certain circumstances the player loses control of his war game troops and is obliged to watch them—being moved by him, of course—do all sorts of things he does not want them to do. This is especially the case with morale tests where, as already pointed out, in adverse circumstances troops might either halt, run away or go into an uncontrolled advance all depending upon what was happening around them, the tactical situation plus what we have come to know as the 'random factor', the imponderable as found from a dice throw. Only rarely will really bad dice throwing substantially upset good war game tactics, and conversely it is infrequent for a series of lucky throws to win a game unsupported by good tactics, but it does represent the element of chance which is omnipresent in warfare. In the case of hand-to-hand combat, for example, a good dice throw can naturally help, but basically the result is determined by the troops involved and the weapons they use. The dice throw can alter the effect only to a limited degree on either side of the 'norm' which has been established through research.

Another important feature of the *Research Group Rules* is the manner in which they set out values for every type of soldier or warrior likely to be employed in an ancient war game, this being based partly on the weaponry of the individual, and partly on the ease or difficulty with which a particular nation was able to produce different troop types. This includes the natural warlike potential of a country as well as the cost to the exchequer or whatever. Achaemenid Persia could furnish masses of light cavalry from the peoples within her borders who were practically born in the saddle and lived by spear and bow, while for a nation like Rome to do likewise was more difficult, their light cavalry having to be recruited and almost invariably trained from scratch—until Rome was able to recruit from subject or allied races, of course. Thus, when we come to the stage of choosing armies for a war game, one Persian the other Roman, it will cost the latter 8 points for a light cavalryman while the Persian gets his for only 6. In war games with equal point armies, this is very important. It has nothing to do with the immediate fighting value of the individuals, but it does ensure the avoidance of armies of unrealistic composition.

Three editions of the *Research Group Rules* have been published, together with occasional amendments and they have now been extended to include the period 1000 BC to 1000 AD, the latter taking them into the early middle ages, but this does not affect the period with which we are concerned, and at the moment the *Rules* are as close to being faultless as one could wish, allowing for the fact that everyone has his own private ideas. It might be said that in my own group, of which I have already spoken, they have been tested as well as they have been anywhere, with one game per week for nearly three years. Occasionally a situation will arise where one has to stop and think but a little thought and consultation invariably finds a solution from the printed page and all goes on. Certainly any improvement could be only trifling and the ancient wargamer has much for which to thank the Research Group.

Appendix IV—The Society of Ancients

No book concerning itself with ancient wargaming could be complete without a reference to the Society of Ancients, the organisation which concerns itself with the study of military history in the ancient and medieval periods and with wargaming in the same epochs. The ancient wargamer is more fortunate than his brethren who favour other periods in that he can be a member of a society which is the only one of its kind that has existed for any length of time and which fosters only one particular aspect of the wargaming hobby. The Society was formed in 1965 and has since that date produced a bi-monthly magazine, *Slingshot*, a totally non-commercial undertaking, which caters exclusively for the members and deals with every fact of military affairs in the relevant periods. The content ranges from the quite technical—the study of ancient military manoeuvres, details of armour and weapons—to the lightest side of wargaming, by way of war game reports, discussion of rules, reports of meetings and other things dear to the heart of the enthusiast.

In addition to the many associations formed through the medium of the journal, the Society holds district meetings regularly throughout the country, notably at Southampton, Worthing, Birmingham and Glasgow. These are weekend events, during which large and small war games take place and members meet for discussion and exchange of ideas. It is indicative of the spread of ancient wargaming to note that similar conventions have taken place in far off California, where there exists a veritable hotbed of ancient wargamers.

From very small beginnings the Society, as might be expected, has increased its membership by leaps and bounds, each year seeing it a minimum of 50% larger than during the previous one and this trend shows no sign of decreasing. For the ancient wargamer as well as the straightforward student of military history, membership of the society is an essential.

Another attraction is the Society war games championship which is organised on a league basis. Members from every part of the country participate as well as many in America and New Zealand.

At the time of writing the annual subscription is a mere £1·50 and interested parties can obtain all information required from the Treasurer, A. L. Nickels, 63 Cottenham Park Road, Wimbledon, London SW20.

Appendix V—
Availability of War Game Figures

The present day ancient wargamer is indeed fortunate in being able to repair to his nearest model or hobby shop and acquire almost any sort of figure he wants, irrespective of type or period. The ancient era is very adequately catered for nowadays and a wide choice is available. Manufacturers have generally settled for the 25mm figure as being the most desirable size for the war game figure, combining as it does handiness of size with enough detail to impart a degree of character, and many of the most recently produced figures are miniature gems, with facial characteristics and equipment as would at one time have been a credit to any 54mm figure. This is fine, of course, but renders the wargamer prey to severe temptations to spend and spend—and who might blame him faced with such desirable figures. Practically every period can be had, from ancient Egypt onwards, although certain countries are more popular than others, this being not unconnected with their potency on the war games table. In assembling basic information of manufacturers' product I make no recommendations as to which are in my view most desirable—each wargamer has his own preferences and I leave the choice to him. It should, however, be pointed out that different makers have varying ideas, it seems, as to what consitutes 25mm, and one must be prepared to find variations of size between products.

For the record, however, the following are the best known firms in production at the time of writing:

Miniature Figurines Ltd., 28/23 Northam Road, Southampton. A truly tremendous range of ancient figures of every race and army, up to medieval Japanese Samurai warriors.

Hinchliffe Models, Meltham, Huddersfield, Yorks. Long famous for magnificent model artillery pieces, this firm now produces model figures, including types for ancient wargaming—Greeks, Persians and Romans. The ancient line is to be greatly increased.

Greenwood and Ball Ltd., 'Martinhoe', East End Way, Pinner, Middlesex. Produces 'Garrison' war game figures. A lengthy list of 'ancients', mainly Greek, Roman and Persian, with some Gauls and Carthaginians. Now brings out magnificent 54mm gladiators and has similar 25mm types on the way.

Phoenix Model Developments, Ltd., The Square, Earls Barton, Northampton. Beginning to produce 25mm Greeks and Persians. Also has the former in 30mm size.

Peter Laing, 11 Bounds Way, Southborough, Tunbridge Wells, Kent. Produces a smaller figure than is usual—15mm. At the time of writing has Greeks and Persians.

Bibliography

The following can be called in all sincerity a selective bibliography and it is so termed because each work quoted has provided valuable reference material to me in connection with ancient wargaming and as such merits my making it known to my fellows. They are all modern works, as I assume that the true enthusiast will seek out the original sources himself for study in conjunction with present day commentaries. They—the originals—are essential in themselves, and in many cases tremendously readable, the histories of Tacitus, Herodotus and Arrian are all fascinating, and some are enormously exciting—who could fail to be thrilled by Xenophon's Anabasis or Caesar's Gallic War. All these and many other contemporary sources can be found in the bibliographies of many of the books I list, but to save space and time, let me include only those modern works which I have found of the greatest value.

F. E. Adcock: *The Greek and Macedonian Art of War* (1957)
The Roman Art of War under the Republic (1940)

K. J. Anderson: *Military Theory and Practice in the Age of Xenophon* (1970)

P. Barker: *Armies of the Macedonian and Punic Wars* (1971)
The Armies and Enemies of Imperial Rome (1972)

J. G. P. Best: *Thracian Peltasts and their Influence on Greek Warfare* (1969)

A. R. Burn: *Persia and the Greeks* (1962)
The Warring States of Greece (1968)

G. de Beer: *Hannibal* (1969)

T. A. Dodge: *Hannibal* (1891)

J. F. C. Fuller: *Decisive Battles of the Western World: Vol. 1* (1939)
The Generalship of Alexander the Great (1958)
Julius Caesar—Man, Soldier and Tyrant (1965)
Peter Green: *The Year of Salamis* (1970)
Armada from Athens (1971)
C. Hignet: *Xerxes' Invasion of Greece* (1923)
E. M. Lloyd: *A Review of the History of Infantry* (1908)
E. W. Marsden: *The Campaign of Gaugamela* (1964)
R. D. Milns: *Alexander the Great* (1963)
Montgomery of Alamein: *A History of Warfare* (1968)
H. W. Parke: *Greek Mercenary Soldiers* (1933)
L. Rossi: *Trajan's Column and the Dacian Wars* (1971)
H. H. Scullard: *Scipio Africanus, Soldier and Statesman* (1950)
A. M. Snodgrass: *Arms and Armour of the Greeks* (1967)
G. R. Watson: *The Roman Soldier* (1969)
G. Webster: *The Roman Imperial Army* (1969)
Y. Yadin: *The Art of Warfare in Biblical Times* (1963)

Index